Cornerstones of Undecidability

Prentice Hall International Series in Computer Science

C. A. R. Hoare, Series Editor

BACKHOUSE, R. C., *Program Construction and Verification*
BACKHOUSE, R. C., *Syntax of Programming Languages*
DEBAKKER, J. W., *Mathematical Theory of Program Correctness*
BARR, M. and WELLS, C., *Category Theory for Computing Science*
BEN-ARI, M., *Principles of Concurrent and Distributed Programming*
BEN-ARI, M., *Mathematical Logic for Computer Science*
BIRD, R. and WADLER, P., *Introduction to Functional Programming*
BORNAT, R., *Programming from First Principles*
BOVET, D. and CRESCENZI, P., *Introduction to the Theory of Complexity*
BUSTARD, D., ELDER, J. and WELSH, J., *Concurrent Program Structures*
CLARK, K. and McCABE, F. G., *Micro-Prolog: Programming in Logic*
CROOKES, D., *Introduction to Programming in Prolog*
DAHL, O.-J., *Verifiable Programming*
DROMEY, R. G., *How to Solve It by Computer*
DUNCAN, E., *Microprocessor Programming and Software Development*
ELDER, J., *Construction of Data Processing Software*
ELLIOTT, R. J. and HOARE, C. A. R. (eds), *Scientific Applications of Multiprocessors*
FREEMAN, T. L. and PHILLIPS, R. C., *Parallel Numerical Algorithms*
GOLDSCHLAGER, L. and LISTER, A., *Computer Science: A modern introduction (2nd edn)*
GORDON, M. J. C., *Programming Language Theory and its Implementation*
GRAY, P. M. D., KULKARNI, K. G. and PATON, N. W., *Object-oriented Databases*
HAYES, I. (ed), *Specification Case Studies (2nd edn)*
HEHNER, E. C. R., *The Logic of Programming*
HENDERSON, P., *Functional Programming: Application and implementation*
HOARE, C. A. R., *Communicating Sequential Processes*
HOARE, C. A. R., and GORDON, M. J. C. (eds), *Mechanized Reasoning and Hardware Design*
HOARE, C. A. R., and JONES, C. B. (eds), *Essays in Computing Science*
HOARE, C. A. R., and SHEPHERDSON, J. C. (eds), *Mechanical Logic and Programming Languages*
HUGHES, J. G., *Database Technology: A software engineering approach*
HUGHES, J. G., *Object-oriented Databases*
INMOS LTD, *Occam 2 Reference Manual*
JACKSON, M. A., *System Development*
JOHNSTON, H., *Learning to Program*
JONES, C. B., *Systematic Software Development Using VDM (2nd edn)*
JONES, C. B. and SHAW, R. C. F. (eds), *Case Studies in Systematic Software Development*
JONES, G., *Programming in Occam*
JONES, G. and GOLDSMITH, M., *Programming in Occam 2*
JONES, N. D., GOMARD, C. K. and SESTOFT, P., *Partial Evaluation and Automatic Program Generation*
JOSEPH, M., PRASAD, V. R. and NATARAJAN, N., *A Multiprocessor Operating System*
KALDEWAIJ, A., *Programming: The derivation of algorithms*
KING, P. J. B., *Computer and Communications Systems Performance Modelling*
LALEMENT, R., *Computation as Logic*
LEW, A., *Computer Science: A mathematical introduction*
McCABE, F. G., *Logic and Objects*
McCABE, F. G., *High-Level Programmer's Guide to the 68000*
MEYER, B., *Introduction to the Theory of Programming Languages*
MEYER, B., *Object-oriented Software Construction*
MILNER, R., *Communication and Concurrency*
MITCHELL, R., *Abstract Data Types and Modula 2*
MORGAN, C., *Programming from Specifications*
PEYTON JONES, S. L., *The Implementation of Functional Programming Languages*
PEYTON JONES, S. and LESTER, D., *Implementing Function Languages*
POMBERGER, G., *Software Engineering and Modula-2*
POTTER, B., SINCLAIR, J. and TILL, D., *An Introduction to Formal Specification and Z*
REYNOLDS, J. C., *The Craft of Programming*
ROSCOE, A. W., *A Classical Mind: Essays in honour of C. A. R. Hoare*
RYDEHEARD, D. E. and BURSTALL, R. M., *Computational Category Theory*
SLOMAN, M. and KRAMER, J., *Distributed Systems and Computer Networks*
SPIVEY, J. M., *The Z. Notation: A reference manual (2nd edn)*
TENNENT, R. D., *Principles of Programming Languages*
TENNENT, R. D., *Semantics of Programming Languages*
WATT, D. A., *Programming Language Concepts and Paradigms*
WATT, D. A., *Programming Language Processors*
WATT, D. A., WICHMANN, B. A. and FINDLAY, W., *ADA: Language and methodology*
WELSH, J. and ELDER, J., *Introduction to Modula 2*
WELSH, J. and ELDER, J., *Introduction to Pascal (3rd edn)*
WELSH, J., ELDER, J. and BUSTARD, D., *Sequential Program Structures*
WELSH, J. and HAY, A., *A Model Implementation of Standard Pascal*
WELSH, J. and McKEAG, M., *Structured System Programming*
WIKSTRÖM, Å., *Functional Programming Using Standard ML*

Cornerstones of Undecidability

Grzegorz Rozenberg
and
Arto Salomaa

Prentice Hall
New York   London   Toronto   Sydney   Tokyo   Singapore

First published 1994 by
Prentice Hall International (UK) Limited
Campus 400, Maylands Avenue
Hemel Hempstead
Hertfordshire, HP2 7EZ
A division of
Simon & Schuster International Group

© Grzegorz Rozenberg and Arto Salomaa 1994

All rights reserved. No part of this publication may be reproduced, stored in a retrieval system, or transmitted, in any form, or by any means, electronic, mechanical, photocopying, recording or otherwise, without prior permission, in writing, from the publisher.
For permission within the United States of America contact Prentice Hall Inc., Englewood Cliffs, NJ 07632

Printed and bound in Great Britain at the
University Press, Cambridge

Library of Congress Cataloging-in-Publication Data

A catalog record for this book is available from
the Library of Congress

ISBN 0-13-297425-8 (pbk)

British Library Cataloguing in Publication Data

A catalogue record for this book is available from
the British Library

ISBN 0-13-297425-8 (pbk)

1 2 3 4 5   98 97 96 95 94

To
Daniel, Suvi, Juhani and Daniel

Contents

Dramatis personae

David	David Hilbert	(1862–1943)
Emil	Emil Post	(1897–1954)
Kurt	Kurt Gödel	(1906–1978)
Tarzan	A traveller	
Bolgani	A guide	

Preface		ix
Prologue		xi
1	**Halting Problem**	**1**
1.1	Recursiveness and Church's Thesis	1
1.2	Recursive enumerability, machine models and complexity	14
1.3	Language theory and word problems	25
1.4	Creative sets and Gödel theory	37
	Interlude 1	46
2	**Post Correspondence Problem**	**50**
2.1	What Emil said	50
2.2	Analyzing PCP further	56
2.3	Equality sets and test sets	65
2.4	Emil's words in retrospect: an overview of PCP	77
3	**Diophantine Problems**	**91**
3.1	Reduction to machine models	91

3.2	Hilbert's Tenth Problem	107
3.3	Power series and plant development	118

Interlude 2 — **127**

4 Classes of Problems and Proofs — **131**
- 4.1 Problems with many faces — 131
- 4.2 Borderline between decidable and undecidable — 146
- 4.3 Mathematical proofs: step by step and zero-knowledge — 152

5 The Secret Number — **161**
- 5.1 Compressibility of information — 161
- 5.2 Randomness and magic bits — 179

Epilogue — **189**

Bibliography — **193**

Index — **195**

Preface

The enormous development of computers and the resulting profound changes in scientific methodology have opened new horizons for the science of mathematics that have no parallel over its long history. One of the very central ideas of modern mathematics is *undecidability*. The nonexistence of a general method to solve all instances of a specific problem and the impossibility of proving or disproving a statement within a formal system are facts that cannot be ignored. They are not exceptional and pathological but are numerous, familiar and palpable.

The issues concerning undecidability lead us far beyond ordinary mathematics. Understanding the power and limitations of human knowledge is an exciting but tremendously difficult task. This task is approached in this book by investigating the cornerstones and key issues of undecidability. Many questions of fundamental philosophical importance will be asked. For instance, in the final chapter our considerations lead to a mysterious number that encodes the cornerstones of undecidability very compactly, a number that can be justly called "the secret number", "the number of wisdom" or "the number that can be known of but not known through human reason". Related questions are, for instance: "How are randomness and uncomputability related?" "Is the universe ordered?" and "Does God play dice?"

The book can be used as a text or as supplementary reading material for courses in theoretical computer science and mathematics, in particular for courses dealing with mathematical logic, formal languages, automata and abstract theory of algorithms. Since the presentation of material and proofs is largely innovative, and includes new results, the book should also be of interest to the specialist as a monograph and as a work for reference. The informal conversations (Prologue, Interludes, Epilogue) make at least parts of the book accessible for the educated layman. The participants in these conversations discuss the "mysteries" of undecidability, the three famous mathematicians using their own words — sometimes in a slightly adapted form.

The informal conversations mentioned also contain much of the customary prefatory material such as historical background, introduction to the area, hints to the reader, and acknowledgements.

Grzegorz Rozenberg
Arto Salomaa
October, 1993

Prologue

David. Everything which can be the object of scientific thinking and is sufficiently mature for theory formation falls within the framework of axiomatic method. By building new and more sophisticated axiomatic structures, we gain deeper insights and more uniform knowledge. Mathematics plays a leading role among all sciences in the use of the axiomatic method.

The *consistency* of the axioms is of utmost importance. The presence of a contradiction in a theory means the collapse of the whole theory. Often, the consistency of a theory is taken to be obvious, although, in reality, involved mathematical arguments would be needed for the proof. It is not sufficient simply to avoid known contradictions by altering the axioms. One must go much further and require that contradictions are not at all possible in the proposed axiomatic system.

This line of research, anticipated by Frege and completed by Russell, leads to the axiomatization of logic and can be considered as the culmination of axiomatization in general. Much work still remains to be done on many fronts. The question about the consistency of the arithmetic of integers is by no means isolated. On the contrary, it belongs to an area of the most difficult epistemological problems having a specific mathematical flavor. To this area belong the problem of the *solvability* (in principle) of every mathematical question, the problem of the *verifiability* of a mathematical result, the problem of a *complexity measure* for mathematical proofs, the problem of the relation between *substance* and *formalism* in mathematics and, finally, the problem of the *decidability* of a mathematical question in finitely many steps.

Tarzan. Decidability — this is exactly what I want to learn more about.

David. Among the problems mentioned, the last one (the problem of decidability in finitely many steps) is the best known and most frequently discussed because it is intimately connected with the essence of mathematical thinking.

Bolgani. Already Leibniz thought in terms of an ideal situation, where mathematicians settle their disputes by computations, using some effective procedure.

(Effective procedures are nowadays customarily referred to as *algorithms*.) When in disagreement, mathematicians should, according to Leibniz, just say "calculemus" ("let us compute") and then apply some algorithm. Nobody ever thought that this would produce solutions on line. The mathematicians might be dead before the algorithm terminates. But, given enough time, it would eventually terminate.

David. We cannot be satisfied with the axiomatization of mathematics and logic before all problems mentioned in this context are properly understood. The decidability of a mathematical question in finitely many steps means that we have reached the same situation as in the task of computing the $10^{10^{10}}$th digit in the decimal expansion of π — the solvability of the question is obvious but the solution remains unknown. Indeed, among the problems mentioned (such as verifiability, complexity measure, substance and formalism), the one concerning decidability creates in my opinion a new research area. To conquer this area we must take the notion of a mathematical proof itself as an object of study, in the same way as an astronomer takes into consideration the motion of the observatory, a physicist is concerned about the theory of his equipment, and a philosopher criticizes his own mind.

Bolgani. Here David describes his program that was to become famous later on.

David. I believe that everything, which can be an object of scientific thinking and is mature enough for the formation of a theory, belongs to the realm of the axiomatic method and in this sense also to mathematics. We become more aware of the unity of our knowledge and gain deeper insights into the essence of scientific thinking by penetrating deeper into the world of axioms. Mathematics is chosen to a leading role in science as the source of the axiomatic method.

Tarzan. We have already heard most of this before. This landscape looks very bright. But I see a thunderstorm approaching.

Kurt. The development of mathematics towards greater precision has led, as is well known, to the formalization of large tracts of it, so that one can prove any theorem using nothing but a few mechanical rules. The most comprehensive formal systems that have hitherto been set up are the system *Principia Mathematica*, on the one hand, and the Zermelo-Fraenkel axiom system of set theory (further developed by J. von Neumann), on the other. These two systems are so comprehensive that in them all methods of proof used today in mathematics are formalized, that is, reduced to a few axioms and rules of inference.

Bolgani. This sounds like [Göd 1], one of the most famous scientific papers in this century. And now comes the blow.

Kurt. One might therefore conjecture that these axioms and rules of inference are sufficient to decide *any* mathematical question that can be at all formally expressed in these systems. It will be shown below that this is not the case, that, on the contrary, there are in the two systems mentioned relatively simple problems

Prologue

in the theory of integers that cannot be decided on the basis of the axioms. This situation is not in any way due to the special nature of the systems that have been set up, but holds for a wide class of formal systems; among these, in particular, are all systems that result from the two just mentioned through the addition of a finite number of axioms, provided no false propositions become provable owing to the added axioms.

Bolgani. Later on the result was expressed in various more explicit ways.

Kurt. For each formal system of mathematics there are propositions which can be *expressed* within this system but which cannot be *decided* by the axioms of this system. These propositions even have a relatively simple form since they belong to the theory of positive integers. The fact that the whole of mathematics cannot be captured in a formal system already follows from Cantor's diagonal argument, but it still remained conceivable that at least certain subsystems of mathematics could be formalized completely. My proof shows that even this is impossible if the subsystem includes at least the notions of multiplication and addition of integers.

Tarzan. I know Cantor's diagonal argument. You try to enumerate all real numbers between 0 and 1, and then define a number whose nth decimal differs from the nth decimal of the nth number in the enumeration. If this holds for all n, the newly defined number is not among the originally enumerated ones. But tell me about *Principia Mathematica* and Zermelo-Fraenkel.

Bolgani. *Principia Mathematica* by Russell and Whitehead is the most comprehensive work ever written about the logical foundations of mathematics. The philosophical thesis was that all the concepts of mathematics can be expressed in logic and all theorems can be proved using only the axioms and rules of inference of logic. In this *logicism* it was impossible to go outside logic; consistency could only mean that no contradiction had been found so far. Gödel had grave doubts about logicism and also about Hilbert's *formalism*, with its preoccupation about consistency. We do not have to worry about the details of the Zermelo-Fraenkel axiomatization. In fact, I do not know the details. Very few people do, although everybody uses the term.

I would like to raise a different issue. There is an old saying, going back to Socrates, according to which a wise man differs from others in knowing what he does not know. In this sense science has entered the phase of wisdom in this century: it has started to discover its own limitations. The best known example is in quantum mechanics, where Heisenberg's uncertainty principle imposes limits on what can be measured. Perhaps even more disconcerting are the incompleteness and undecidability results in mathematics. Let us hear some of the first reactions.

David. The final goal is to know that our customary methods in mathematics are totally consistent. Concerning this goal, I would like to stress that the view, temporarily widespread — that certain recent results of Gödel's imply that my proof theory is not feasible — has turned out to be erroneous. In fact, those results show only that, in order to obtain an adequate proof of consistency, one must use the finitary standpoint in a sharper way than is necessary in treating the elementary formalism.

Bolgani. The spread of the view was not temporary, but has continued ever since. Hilbert had in mind something that was no more finitary and leads to formal systems where, for instance, theorems cannot be effectively enumerated. I would like to return to this matter later on.

Emil. The following remarks should prove significant in the development of symbolic logic along the lines of Gödel's theorem on the incompleteness and Church's results concerning absolutely unsolvable problems. We have in mind a *general problem* consisting of a class of *specific problems*. A solution of the general problem will then be one that furnishes an answer to each specific problem.

Bolgani. That's right. When we speak of a *decision problem DP* in the sense of Hilbert and want to construct an algorithm for its solution, then DP consists of many, in fact of infinitely many, specific instances. We ask many questions such as "Is a given integer n prime?" A specific instance results by considering a specific value of n. For finitely many instances we can, at least in principle, construct a table of answers and consider the table look-up as the required algorithm.

Tarzan. I would be happy to see your table of answers if I give you questions like "How many species of mollusks inhabit the earth today at noon?" or "What will the recorded number of species of flatworms be in the year 2100?" But returning to the more serious line, I could hear two important issues mentioned. The first one was *incompleteness*, which I understand means that certain statements are independent of a formal system, that is, the statement can neither be proved nor disproved within the system. The second one was *absolute unsolvability* or *absolute undecidability*, the nonexistence of an algorithm.

Bolgani. The two issues are interrelated, as we will see. To speak of the nonexistence of a specific algorithm or a specific mollusk, we have to agree what algorithms or mollusks are. The customary definition of an algorithm is in terms of a Turing machine. Let us hear a very simply formulated contemporary definition from the mid-1930s [Post 1].

Emil. Two concepts are involved: that of a *symbol space* in which the work leading from problem to answer is to be carried out, and a fixed unalterable *set of directions* which will both direct operations in the symbol space and determine the order in which those directions are to be applied.

The symbol space is to consist of a two-way infinite sequence of spaces or boxes. The problem solver or worker is to move and work in this symbol space, being capable of being in, and operating in but one box at a time. And apart from the presence of the worker, a box is to admit but two possible conditions, i.e., being empty or unmarked, and having a single mark on it, say, a vertical stroke.

One box is to be singled out and called the starting point. We now further assume that a specific problem is to be given in symbolic form by a finite number

Prologue

of boxes being marked with a stroke. Likewise the answer is to be given in symbolic form by such a configuration of marked boxes. To be specific, the answer is to be the configuration of marked boxes left at the conclusion of the solving process.

The worker is assumed to be capable of performing the following primitive acts, as well as otherwise following the directions described below:

(a) Marking the box he is in,
(b) Erasing the mark in the box he is in,
(c) Moving to the box on his right,
(d) Moving to the box on his left,
(e) Determining whether the box he is in, is or is not marked.

The set of directions which, be it noted, is the same for all specific problems and thus corresponds to the general problem, is to be of the following form. It is to be headed:
Start at the starting point and follow direction 1. It is then to consist of a finite number of directions to be numbered $1, 2, \ldots, n$. The ith direction is to have one of the following forms:

(**A**) Perform operation $0_i [0_i = (a), (b), (c)$ or $(d)]$ and then follow direction j_i,
(**B**) Perform operation (e) and, according as the answer is yes or no, correspondingly follow direction j_i' or j_i'',
(**C**) Stop.

Clearly one direction only needs to be of type C. Note also that the state of the symbol space directly affects the process only through directions of type B.

A set of directions applicable to a general problem sets up a deterministic process when applied to each specific problem. This process will terminate when and only when it comes to the direction of type (C). The set of directions will then be said to set up a *finite 1-process* in connection with the general problem *if the process it determines terminates for each specific problem*. A finite 1-process associated with a general problem will be said to be a *1-solution* of the problem if the answer it thus yields for each specific problem is always correct.

Tarzan. That was no small talk but I could still follow it. I understand we got a definition of decidability. A general problem is decidable if it has a 1-solution. Otherwise, it is absolutely undecidable.

Bolgani. You are right: a 1-process is a formalization of the notion of an algorithm or effective procedure. The terms "1-process" and "1-solution" never came into common usage. The description we heard is compact and precise. A specific problem is encoded in the symbol space by marking some boxes. It is interesting to observe that the "worker" carrying out the directions can be visualized as an industrious clerk or a machine. The words Emil uses are neutral in this respect. Turing spoke of a "machine capable of a finite number of conditions". We will make further comparisons later. Now let us hear how the ideas of incompleteness and undecidability can be brought together.

Emil. With some modification the above formulation is also applicable to symbolic logics. We do not now have a class of specific problems but a single initial finite marking of the symbol space to symbolize the primitive formal assertions of the logic. On the other hand, there will now be no direction of type (C). Consequently, a deterministic process will be set up which is *unending*. We further assume that in the course of this process certain recognizable symbol groups, i.e., finite sequences of marked and unmarked boxes, will appear which are not altered further in the course of the process. These will be the derived assertions of the logic. Of course, the set of directions corresponds to the deductive processes of the logic. The logic may then be said to be 1-*generated*.

An alternative procedure, less in keeping, however, with the spirit of symbolic logic, would be to set up a finite 1-process which would yield the nth theorem or formal assertion of the logic, given n.

Tarzan. No matter how we set up the logic, there will be statements about arithmetic such that neither the statement itself nor its negation ever appears in the generated list. Unless the logic is inconsistent and, consequently, every statement appears in the list. I start to understand vaguely. But there are many details that are not clear. I also have millions of questions. I would like to learn more. Discussions can go on indefinitely. To gain a deeper understanding, one has to go into some technicalities.

Bolgani. I have just completed a mathematical text about undecidability, five chapters. It begins with the basics but then goes into more sophisticated matters, also some recently discovered ones. You should read it.

Tarzan. What are the prerequisites on the part of the reader? Will I understand it?

Bolgani. You should. Some parts require careful study and thinking. It is helpful if the reader has already visited the world of Turing machines and computability. Some basic mathematical notions, such as congruences and matrices, are assumed to be known. Really, not much factual knowledge is required.

Tarzan. Then to work.

Chapter 1

Halting Problem

1.1 Recursiveness and Church's Thesis

The starting point for our discussions is the informal notion of a *computable* function, as well as the informal notion of a *decidable* problem. We call a function f (*effectively*) *computable* when there exists an *effective procedure* (or *algorithm*; for us the two terms are synonymous) that produces the value of f correctly for each possible argument. Let us analyze further what this means.

We consider functions $f(x_1, \ldots, x_k)$, $k \geq 1$, whose variables x_i range over nonnegative integers and whose values are also nonnegative integers. For all our purposes, we may restrict the domain and range of f in this fashion; this situation can always be reached by a suitable encoding of the objects we are considering. Such an encoding is rather straightforward if the objects are words over a certain alphabet but more involved if the objects are, for instance, theorems in a formal system, and we want somehow to exhibit the property of being a theorem in the encoding. We allow also the possibility of the function f being *partial*: the domain of f does not necessarily consist of all k-tuples $(x_1, \ldots, x_k)$ of nonnegative integers but the value $f(x_1, \ldots, x_k)$ may be undefined for some arguments $(x_1, \ldots, x_k)$. The term *total* is used if there are no such argument values.

An effective procedure A_f for computing f must meet the following two conditions.

(i) The description of A_f must be finite in length and sufficiently detailed that no ingenuity is required on the part of the person or machine executing A_f. The execution of A_f is only a matter of following the instructions provided in the description of A_f carefully.

(ii) If the k-tuple $(x_1, \ldots, x_k)$ is in the domain of f then, after the execution of finitely many discrete computation steps, the procedure A_f halts and produces the value $f(x_1, \ldots, x_k)$. If $(x_1, \ldots, x_k)$ is not in the domain of f, then A_f may produce an answer saying so, get stuck at some stage, or might run forever, never halting. However, in this case (ii), A_f never stops producing a nonnegative integer as an alleged value $f(x_1, \ldots, x_k)$.

The execution of A_f may be depicted as a computer, with an unbounded memory and no time restrictions, following a program. Before the advent of computers, the intuitive picture was that of an industrious tireless clerk following instructions carefully.

Example 1.1. All basic arithmetic functions are effectively computable. For instance, let $f(0) = 1$ and let $f(x)$, $x \geq 1$, be the xth prime in order of magnitude. Clearly, the primality of any number n can be tested effectively. A classical method for this is the *Sieve of Eratosthenes*: test the divisibility of n by all integers less than or equal to $\sqrt{n}$. An effective procedure A_f for computing $f(x)$ consists of the following recursion. Initially, we know that $f(1) = 2$ and $f(2) = 3$. If we have already computed $f(x)$, $x \geq 2$, then $f(x+1)$ is the first prime among the numbers $f(x) + 2$, $f(x) + 4$, $f(x) + 6$, ... Since there are infinitely many primes, we always find the value $f(x+1)$ by an effective procedure, which means that the function f is total. The procedure A_f can be described more formally in a chosen programming language.

□

Restrictions (i) and (ii) imposed on computable functions do not imply any restrictions concerning the efficiency of computations: effectiveness should be kept apart from efficiency. Thus, although we require that the effective procedure A_f produces the value $f(x_1, \ldots, x_k)$ after finitely many computation steps, no upper bound is imposed in advance on the number of steps. Similarly, no upper bound is imposed in advance on the amount of memory space needed to carry out the procedure A_f for a given argument. Issues concerning time and space requirements are dealt with in the theory of *computational complexity*. The theory will be discussed briefly in Section 1.2.

Thus, effective computability does not mean "practical" computability. We may agree that something is not practically computable if it cannot be computed in a lifetime using the best available computing devices. Consider the effectively computable function $f(x)$ from Example 1.1. The value $f(10^{10})$ is practically computable, whereas $f(10^{10^{10}})$ will probably never become practically computable. As regards effectively computable functions, there are no practical restrictions concerning time and memory space. In this sense, the class of effectively computable functions constitutes an upper bound on what can ever be considered practically computable.

The informal notion of a *decidable problem* can be reduced to that of a computable function as follows. All problems that we are going to consider have a "yes" or "no" answer. The problem can be identified by the set of its "yes" instances. The set of "yes" instances of the problem "Is n prime?" is the set of prime numbers. Instead of deciding the problem itself, we may decide membership in the sets of "yes" instances. This again amounts to computing the *characteristic function* $f_S(n)$ of the set S:

$$f_S(n) = \begin{cases} 1 & \text{if } n \text{ is in } S, \\ 0 & \text{if } n \text{ is not in } S. \end{cases}$$

Accordingly, we say that a problem is (effectively) *decidable* iff the characteristic function of its "yes" instances is (effectively) computable. If the set of "yes" instances is not a set of nonnegative integers, it is first encoded as such a set.

Example 1.2. Consider words over the alphabet $\Sigma = \{a, b\}$, that is, finite sequences of letters a and b. We say that a word w is *cube-free* iff w cannot be written as $w = yxxxz$. Here x, y, z are words, and y and z are allowed to be empty. For instance, the words

abbabaabbaababba and *abbabbaba*

are cube-free, whereas the words

aaa, *abbabaabaababbab* and *bbabababa*

are not cube-free. Our problem is to decide whether or not an arbitrary word w consisting of the letters a and b is cube-free. Thus, the first two words given above belong to the set of "yes" instances of our problem.

It is fairly obvious how we can view our problem as a problem of computing values of a function $f(x)$ from nonnegative integers into nonnegative integers. We first view words as nonnegative integers in the so-called *dyadic* representation: the letters a and b become the digits 1 and 2, respectively, and a word $c_0 c_1 \ldots c_n$ becomes the number

$$c_0 2^n + c_1 2^{n-1} + \cdots + c_n.$$

It is easy to see that in this fashion a one-to-one correspondence is established between words over the alphabet $\{a, b\}$ and nonnegative integers. (In this correspondence, the number 0 is assigned to the empty word.) Some illustrations are given in the following table, where the numbers are in decimal notation.

word	number
ab	4
ba	5
aaa	7
bbb	14
abbabbaba	729
bbababab	468

For instance, $729 = 2^8 + 2 \cdot 2^7 + 2 \cdot 2^6 + 2^5 + 2 \cdot 2^4 + 2 \cdot 2^3 + 2^2 + 2 \cdot 2 + 1$.

Finally, we define f to be the characteristic function of the set of "yes" instances of our problem:

$$f(n) = \begin{cases} 1 & \text{if } n \text{ corresponds to a cube-free word}, \\ 0 & \text{otherwise}. \end{cases}$$

Thus, considering the table above,

$$f(4) = f(5) = f(729) = 1 \text{ and } f(7) = f(14) = f(468) = 0.$$

Clearly, $f(n)$ is effectively computable: we can find out for all cubes xxx up to a certain length whether or not they appear as a subword in w, the word corresponding to n. Algorithms more efficient than such an exhaustive search can be devised for this problem.

□

It should be clear that if two functions agree for all but finitely many arguments and one of them is effectively computable, then so is the other. The effective procedure for computing one of the functions can be supplemented with a table look-up as regards the disparate values. In this fashion, an algorithm for computing the other function is obtained. This is only one simple observation concerning effectively computable functions. In general, the purpose of computability theory is first to make precise the intuitive idea and then study properties of effectively computable functions. While the positive aspects of the theory, that is, proofs concerning the effective computability of certain functions can be based on the intuitive notion, the same does not hold true as regards the negative aspects. To prove that a function is *noncomputable* or that a problem is *undecidable*, we need a formalized notion of an effective procedure. Otherwise, we cannot show that no effective procedure can possibly exist for the task we are considering.

A terminological remark is in order before we start the formalization. It is customary to use many terms synonymously with the terms "computable" and "decidable", and we will sometimes follow this practice in the following. Instead of decidable problems, one speaks of *solvable* problems. Moreover, the additional attributes *effectively, recursively, algorithmically* do not alter the meaning. The same also applies to results expressing negative results. Thus, we regard all the following terms as synonyms: undecidable, unsolvable, recursively undecidable, recursively unsolvable, algorithmically undecidable, algorithmically unsolvable, effectively undecidable, effectively unsolvable. In classical mathematics decidability is often referred to as *recursiveness*.

We will first formalize computability (or recursiveness) in terms of an imaginary computing machine, named the *Turing machine* after its inventor Alan Turing. The reader is encouraged to read the beginning of the original article by Turing [Tur] in order to learn his reasons for regarding this model as a general model of computers — before the actual advent of physical computers.

A Turing machine can be viewed as a black box, provided with a read–write head. The latter scans a potentially infinite tape that is marked off into squares, each square containing either the blank B or a letter from a fixed alphabet. At any time the read–write head scans one square. The tape being potentially infinite (in both directions) means that the read–write head never reaches the end of it. However, at any time only finitely many squares can be nonblank. A Turing machine operates in discrete time. At each moment in time it is in a specific internal (memory) *state*,

Recursiveness and Church's Thesis

the number of all possible states being finite. Among the states is a specific *initial* state q_0 and a specific *final* state q_F.

To define a Turing machine TM, we have to specify the state set, the alphabet and finitely many *instructions* of the form

$$(q, a) \text{ yields } (q', a', m),$$

where the qs are states, as are letters (possibly B) and m ("move") assumes one of the two values L (left) or R (right). The instruction means that if TM is in state q and scans a square containing a, then it replaces a by a' (possibly $a = a'$), goes to the state q' (possibly $q = q'$), and moves the read–write head one square to the left or right according to m. The following diagram illustrates how the instruction is carried out with $m = L$:

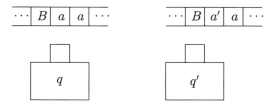

To continue, we would have to see what the instruction is for the pair (q', B). We assume that, for each pair, there is at most one instruction. (This requirement is usually expressed by saying that TM is *deterministic*: at any stage it has at most one possibility for continuation.)

The tape of a Turing machine is used both for input and output, as well as for memory. We now define how a Turing machine computes values of functions $f(x_1, \ldots, x_k)$ considered above. Thus, the variables range over some nonnegative integers (that is, the function need not be defined for all k-tuples) and the values are nonnegative integers as well. We represent an integer n by a sequence of $n + 1$ 1s. (This *unary* notation is the simplest possible. The choice of unary notation, rather than some other notation such as binary or decimal, is irrelevant because a subroutine can be designed for a Turing machine changing one notation to another.) We denote the blank B by 0 in this context. Thus, each tape square contains either 0 or 1. We use $n + 1$ rather than n 1s to represent the number n to distinguish between the blank symbol and the number 0. The notation 1^{n+1} is used to mean $n + 1$ 1s.

The argument $(x_1, x_2, \ldots, x_k)$ is written as

$$1^{x_1+1} 0 1^{x_2+1} 0 \ldots 0 1^{x_k+1}$$

on consecutive squares of the tape. Initially, TM scans the leftmost of these squares in the state q_0, the rest of the tape being blank. A (partial) function $f(x_1, \ldots, x_k)$ is termed *computable* or *partial recursive* iff the following conditions (i) and (ii) are satisfied for some TM:

(i) If $f(x_1, \ldots, x_k)$ is defined, then TM eventually halts in the final state q_F scanning the leftmost 1 in a sequence consisting of $f(x_1, \ldots, x_k) + 1$ 1s, and with the tape blank to the right of this sequence. (Halting in s_F means that there is no instruction for the pair $(q_F, 1)$.)

(ii) If $f(x_1, \ldots, x_k)$ is undefined, then TM never reaches the final state q_F.

If $f(x_1, \ldots, x_k)$ is total, that is, $f(x_1, \ldots, x_k)$ is defined for all k-tuples of nonnegative integers, and some TM satisfies (i) and (ii), then we say that the function f is *recursive* or *total recursive*.

In the above definitions of a Turing machine and a partial recursive function we have not presented all the details formally. However, the remaining informal ideas are easily replaced by formal ones. We just briefly indicate how this can be done. Moreover, in Section 1.3 we will carry out the formalization in terms of rewriting systems.

A *Turing machine* TM is a function whose domain is included in the product set

$$\{0, 1, \ldots, n\} \times \{0, 1\}$$

and whose range is included in the product set

$$\{0, 1, \ldots, n\} \times \{0, 1\} \times \{L, R\}.$$

The intended meaning should be clear. The elements of $\{0, 1, \ldots, n\}$ are the internal states, the first and last being the initial and final states, respectively. The equation $TM(2, 1) = (0, L, 1)$ means that, after scanning 1 in state 2, TM replaces the scanned 1 by 0, moves one square to the left and goes to state 1.

The *computation* of a Turing machine is formalized by using the notion of an *instantaneous description*. The latter is a word over the alphabet $\{0, 1, q\}$ containing exactly two occurrences of q. It is intended to tell the contents of the tape, the position of the read–write head and the state TM is in at some specific time instant. For instance, the instantaneous description $10q11q101$ indicates that the nonblank part of the tape is 10101, and TM scans the second 1 in state 2. If we are computing the value $f(2, 3)$, then the first instantaneous description is $qq11101111$. For $n = 3$, that is, for 3 being the final state, the instantaneous description $10q111q1111100$ indicates that the computation has halted with output 4. If $TM(2, 1) = (0, L, 1)$ and one instantaneous description is $10q11q101$, then the next one is $1q1q0001$. How one instantaneous description leads to another one is defined by cases according to the instructions of the TM.

We can now also give a formal definition concerning the recursiveness of sets S of nonnegative integers. By definition, such an S is *recursive* iff its characteristic function is recursive. Observe that the characteristic function is total. Thus, a problem being *decidable* means that the set of its "yes" instances is recursive.

Recursiveness and Church's Thesis

Before we discuss the adequacy of the formal notion of recursiveness for the intuitive notion of effective computability, we give a simple specific example of a function that is not recursive. Indeed, our example function grows faster than any recursive function. The construction is customarily referred to as the "busy beaver", and can also be used as a simple counterexample to the claim sometimes made: whenever a function has been defined precisely, then it is also effectively computable.

We denote by $T(n)$, $n \geq 1$, the following set of Turing machines. A Turing machine TM with the state set $\{0, 1, \ldots, n\}$ belongs to $T(n)$ iff TM halts in the final state n when started on a blank tape. (Using the notation for instantaneous descriptions described above, we initially have $qq0$. It is assumed that there are no instructions for the state n.) For each TM in $T(n)$, we denote by $ONES(TM)$ the number of 1s on the tape when TM halts. Finally, define

$$BEAVER(n) = \max\{ONES(TM) | TM \text{ in } T(n)\}.$$

Thus, $BEAVER(n)$ equals the maximal number of 1s that an $(n+1)$-state Turing machine as described above can print when started on a blank tape. Clearly, $T(n)$ is finite and $ONES(TM)$ is a specific number for each TM in $T(n)$. Thus, to obtain $BEAVER(n)$, we only have to take the greatest among finitely many numbers. Clearly, $BEAVER$ is a total function.

Theorem 1.1. For any total recursive function $f(x)$, the inequality

$$f(x) < BEAVER(x)$$

is satisfied for all sufficiently large x. Hence, the function $BEAVER(x)$ is not recursive.

Proof. Given a total recursive function $f(x)$, we consider the function $g(x)$ defined by

$$g(x) = \max(f(2x+2), f(2x+3)).$$

Clearly, $g(x)$ is total. It is also a straightforward task in programming to transform the Turing machine computing $f(x)$ into a Turing machine computing $g(x)$. (Although straightforward, this task is still tedious if all details are carried out carefully!) Hence, $g(x)$ is recursive and is computed by a Turing machine TM with, say, m states.

For $x = 0, 1, \ldots$, we now construct a Turing machine TM_x such that TM_x (i) first prints $x+1$ 1s when started on a blank tape, and (ii) then simulates TM, initially scanning the leftmost 1. Clearly, $n = m + x + 2$ is an adequate number of states for TM_x. When started on a blank tape, TM_x halts with at least $g(x) + 1$ 1s on the tape. By the definitions of $BEAVER$ and g, we obtain the following inequalities:

(∗) $f(2x+2) < BEAVER(m+x+2), f(2x+3) < BEAVER(m+x+2).$

Assume now that $k \geq m$. Since $BEAVER$ is clearly monotonic, we obtain by (∗)

$$f(2k+2) < BEAVER(2k+2), \ f(2k+3) < BEAVER(2k+3).$$

But this means that the inequality claimed in Theorem 1.1 holds whenever $x \geq 2m+2$.

□

Example 1.3. It is easy to see that $BEAVER(1) = 1$. Denote the states by q_0 and q_1. When scanning 0 in q_0, TM must go to the state q_1 because, otherwise, it does not halt. (In the formalism given above, L and R are the only possible moves and, thus, in state q_0 TM would encounter the original situation again.) Hence, the best TM can do is to replace the original 0 by 1.

Next consider $BEAVER(2)$. Define a Turing machine TM by listing the instructions:

$$\begin{array}{lll} (q_0, 0) & \text{yields} & (q_1, 1, R), \\ (q_0, 1) & \text{yields} & (q_1, 1, L), \\ (q_1, 0) & \text{yields} & (q_0, 1, L), \\ (q_1, 1) & \text{yields} & (q_2, 1, L). \end{array}$$

(Thus, q_2 is the halt state with no instructions.) Let us see what happens when TM is started on a blank tape. We obtain the following sequence of instantaneous descriptions, where we write q_0, q_1, q_2 instead of our earlier $qq, q1q, q11q$:

$$q_0 0 \ , \ 1q_1 0 \ , \ q_0 11 \ , \ q_1 011 \ , \ q_0 0111 \ ,$$

$$1q_1 111 \ , \ q_2 1111 \ .$$

Hence, $BEAVER(2) \geq 4$. In this case it is easy to carry out an exhaustive analysis and prove that actually $BEAVER(2) = 4$. The values

$$BEAVER(3) = 6 \text{ and } BEAVER(4) = 13$$

are known. For greater argument values, only lower bounds have been given. For instance, $BEAVER(8) > 10^{44}$.

□

We started our discussion with the intuitive notion of an algorithm or effective procedure, as well as with the resulting, also informal, notions of an effectively computable function and of a decidable problem. We then introduced the formal notion of a Turing machine and the resulting formal notion of a partial recursive function and a recursive function. For problems encodable as the membership problem for a set of nonnegative integers, this approach also defines formally what it means for a problem to be decidable. Of course, this approach yields the formal definitions for the corresponding negative notions as well (an undecidable problem, etc.).

Clearly, a Turing machine defines an effective procedure in the intuitive sense. This means that our formal notions are not too broad: everything decidable in the formal sense will also be decidable in the intuitive sense. But are the formal notions perhaps too narrow? Is there something effectively computable in the intuitive sense that cannot be computed by a Turing machine? We claim that the answer to these questions is negative.

The statement that a Turing machine is an adequate and a general-enough formal model for the intuitive notion of an effective procedure is customarily referred to as *Church's Thesis*. Thus, Church's Thesis asserts that the formalization is not too narrow, implying that it is exactly the right one.

In order to prove Church's Thesis, we would have to compare effective procedures (an intuitive notion) and Turing machines (a formal notion). To do this, we would have to formalize the notion of an effective procedure. But then we would have to face the problem: is the introduced formalization equivalent to the intuitive notion? The solution of the problem would require a claim corresponding to Church's Thesis, and so we would end up with an infinite regression.

However, the evidence supporting Church's Thesis is overwhelming. The arguments can be divided into the following three main lines (i)–(iii).

(i) All suggested, very diverse, alternative formalizations of the class of effective procedures have turned out to be (provably) equivalent to the Turing machine formalization. These alternative formalizations include the λ-definable functions by Church, the general recursive functions by Kleene, rewriting systems, grammars, Markov algorithms, Post systems and L systems. We will meet some of these alternative formalizations in what follows. Also, machine models other than Turing machines have been presented as equivalent alternative formalizations. Among such models, *register machines* will be important for our purposes.

(ii) The class of mappings computable by Turing machines has very strong invariance properties. Many modifications, variations and extensions of Turing machines have been proved to be equivalent to the original Turing machine model. Such proofs usually consist of straightforward simulations, and we will use different variations of Turing machines rather freely in what follows. The extensions of Turing machines also include machines with more than one tape and/or read–write head, as well as machines working nondeterministically or in dimensions higher than one. The invariance properties correspond to the intuitive idea that if we have effective procedures for solving certain problems, then we also have effective procedures for problems resulting from the previous ones by "combining" them in a reasonably effective way. Thus, if two functions $f_1(x)$ and $f_2(x)$ are effectively computable, then so is their product $f_1(x) \cdot f_2(x)$ and their composition $f_1(f_2(x))$.

(iii) All intuitively effective procedures (for computing functions and solving problems) considered so far have found their formal counterpart; that is, their "programming" on Turing machines has turned out to be possible. This fact was referred to by Markov as the "normalization principle" of algorithms. Of course, Church's Thesis amounts to extending the normalization principle to all intuitively effective procedures.

In what follows we will use Church's Thesis as an essential tool in proofs. In this way we avoid the really tedious constructions involving the formal Turing machine model. So it is not our intention to start proving lemmas and theorems based on formal definitions. Such a formal development could be simplified by establishing auxiliary results concerning subroutines and special programming techniques. Even such simplifications do not prevent the formal development from becoming too time- and effort-consuming when the discussion goes beyond the trivial initial stages. Indeed, in all existing texts such a formal development is sooner or later (usually sooner!) replaced by arguments based on Church's Thesis.

Formal constructions dealing with Turing machines resemble programming in machine language during the time of early computers. Every memory location (that is, a state of a Turing machine) has a specific number attached to it, and the programming has to take care of the requirement that proper numbers are used in connection with each instruction. This is to be contrasted with the much more pleasant programming in some high-level language. For Turing machines, such a high-level language is based on Church's Thesis. Whenever we have a procedure that is effective in the intuitive sense for some job, we can also construct a Turing machine for the same job. This is a proof method that we will frequently follow in what follows.

We consider the enumeration

$$TM_0, \ TM_1, \ TM_2, \ \ldots$$

of all Turing machines. Such an enumeration can be obtained, for instance, as follows. Every Turing machine, TM, is completely determined by listing the states, the tape alphabet (for future purposes we allow more letters than just 0 and 1 here) and the instructions. If the different items are separated by some marker symbol, such a listing amounts to a word $w(TM)$ that is finite in length. Such words are then enumerated according to length, and according to alphabetic order for words of the same length. If $w(TM)$ is the ith word in this enumeration, we let TM be the ith machine.

From now on we assume the enumeration to be fixed. The number i is referred to as the *index*, or the *Gödel number*, of the Turing machine TM_i. The actual choice of the enumeration is irrelevant. The only requirement is that there are effective procedures for determining TM_i, given i, and for determining an i such that $TM_i = TM$, given TM. In other words, there are effective procedures for going from the index to the machine and from the machine to the index. Let $f_i(x)$ be the partial recursive function of one variable computed by the machine TM_i. (The format of inputs and outputs can be the one described above. Thus, initially, TM_i scans the left end of a sequence of $x+1$ 1s. If TM_i halts with x, then it finally scans the left end of a sequence of $f_i(x) + 1$ 1s.) The number i is also referred to as *an index of the function* $f_i(x)$. Clearly, every partial recursive function has infinitely many indices. This follows because to every Turing machine computing a specific function, we may add another dummy state, which is never reachable during a computation. Then we obtain another Turing machine computing the same function. Thus, the same function can be computed in many ways, some of which are essentially the same.

We are now ready for the basic undecidability result, the *undecidability of the halting problem for Turing machines*. The halting problem consists of deciding whether or not a given Turing machine TM_i halts for a given input x. We express halting also by saying that $TM_i(x)$ (or $f_i(x)$ if we talk about functions) *converges*, as opposed to that it *diverges*. Let us now discuss in detail how we can formalize the decidability and undecidability of the halting problem.

What does an algorithm for solving the halting problem look like? Roughly, the input is the pair (i, x). The output is "yes" or "no", depending on whether $TM_i(x)$ converges or diverges.

It is more convenient to talk about one-place functions. Therefore, we have to establish a one-to-one correspondence between pairs (i, x) and nonnegative integers y. This can be done by using a *pairing function*, for instance the function $\varphi(i, x)$ defined by the table

	0	1	2	3	4	5	...
0	0	1	3	6	10	15	...
1	2	4	7	11	...		
2	5	8	12	...			
3	9	13	⋮				
4	14	⋮					
⋮	⋮						

Thus, values are plotted on the table in the increasing order of magnitude along the diagonals. The function φ can also be defined by the quadratic expression

$$\varphi(i, x) = \frac{1}{2}(i + x + 1)(i + x) + i \, .$$

The function φ establishes a one-to-one correspondence between pairs (i, x) and nonnegative integers. Hence, there are "inverse components" $\varphi_1(x)$ and $\varphi_2(x)$ such that the equation

$$\varphi(\varphi_1(x), \varphi_2(x)) = x$$

holds for all x. These observations are immediate from the table representation of φ. In particular, $\varphi_1(x)$ (resp. $\varphi_2(x)$) is the index of the row (resp. column) in which x lies. Thus,

$$\varphi_1(15) = 0 \, , \quad \varphi_2(15) = 5 \text{ because } \varphi(0, 5) = 15 \, .$$

The table representation also immediately yields an effective procedure for computing the values $\varphi(i, x), \varphi_1(x), \varphi_2(x)$. Hence, by Church's Thesis, these functions are recursive.

We are now ready to define formally what is meant by the decidability of the halting problem for Turing machines. By definition, the *halting problem is decidable* iff the set

$$\{\varphi(i,x) | TM_i(x) \text{ converges}\}$$

is recursive. In other words, the *halting problem is undecidable* iff the function g defined by

$$g(\varphi(i,x)) = \begin{cases} 1 & \text{if } TM_i(x) \text{ converges} \\ 0 & \text{if } TM_i(x) \text{ diverges} \end{cases}$$

is not recursive.

Theorem 1.2. *The halting problem is undecidable.*

Proof. Assume the contrary: g is recursive. Then also the two-place function g_1 defined by $g_1(i,x) = g(\varphi(i,x))$ is recursive: a Turing machine for g_1 first transforms the argument (i,x) to $\varphi(i,x)$ and then simulates the machine for g. Moreover, the function g_2 defined by

$$g_2(x) = \begin{cases} 0 & \text{if } g_1(x,x) = 0 \\ \text{undefined} & \text{otherwise} \end{cases}$$

is partial recursive. Indeed, a Turing machine computing g_1 is easily transformed into a Turing machine computing g_2: input x is first doubled, and then the machine computing g_1 is called. Output 0 for the latter is kept unaltered, whereas a never-ending computation is started from output 1 for the machine computing g_1. (For instance, such a never-ending computation results if a new state is entered such that a perpetual move to the right results.)

Now let j be an index for g_2. We ask whether $TM_j(j)$ converges or diverges. Assume firstly that $TM_j(j)$ converges. Then

$$g(\varphi(j,j)) = g_1(j,j) = 1$$

and, hence, $g_2(j)$ is undefined, which means that $TM_j(j)$ diverges. Assume, secondly, that $TM_j(j)$ diverges. Then

$$g(\varphi(j,j)) = g_1(j,j) = 0$$

and, hence, $TM_j(j)$ converges. Thus, a contradiction results in both cases. This means that our original assumption was wrong, and g is not recursive. □

Theorem 1.2 is the very basic undecidability result. The proof immediately yields the following corollary.

Theorem 1.3. *The function $h(x)$ defined by*

$$h(x) = \begin{cases} 1 & \text{if } TM_x(x) \text{ converges} \\ 0 & \text{if } TM_x(x) \text{ diverges} \end{cases}$$

is not recursive.

Proof. The contradiction in the proof of Theorem 1.2 was based only on pairs (i, x) with $i = x$. In other words, the problem studied was restricted to the special case: does the Turing machine TM_x halt for input x? □

The latter problem is usually referred to as the *self-applicability problem for Turing machines*. A Turing machine TM_x is started with its own index x. Does the computation halt? Theorem 1.3 says that the self-applicability problem is undecidable.

The argument showing the undecidability of the halting problem is "diagonal" in nature: pairs (i, x) with $i = x$ play a crucial role. In fact, in any formalization of the intuitive notion of an effective procedure we face the following *dilemma of diagonalization*.

Assume that we have somehow defined the notion EP of an effective procedure formally. (Here we consider effective procedures for computing mappings of the set of nonnegative integers into the same set.) Each specific instance of an EP possesses a finitary description. By enumerating these descriptions, we obtain an enumeration EP_0, EP_1, ... of all effective procedures. Denote by $EP_i(j)$ the value of the function computed by EP_i, for the argument value j. Consider now the function

$$f(x) = EP_x(x) + 1\,.$$

The following is an algorithm (in the intuitive sense) to compute $f(x)$. Given an input x, start the algorithm EP_x with input x, and add 1 to the output. Thus, if our EP-formalization is adequate, the algorithm to compute $f(x)$ has an index, say t. Hence, for all x,

$$f(x) = EP_t(x)\,.$$

The choice $x = t$ now leads to a contradiction.

This dilemma always arises; we did not specify EP in any way. A way out of the dilemma is to include *partial* functions among the functions computed by effective procedures. Then no contradiction arises because the value $EP_t(t)$ is not necessarily defined. An idea resembling the dilemma of diagonalization is also used in the proofs of Theorems 1.2 and 1.3.

We will be very concerned with proofs of undecidability. In general, they are either *direct* or *indirect*. Direct proofs, such as that of Theorem 1.2, make use of some diagonalization argument. Indirect proofs have the following format. We know that a problem P is undecidable. To show the undecidability of another problem P', we prove that if P' were, in fact, decidable, then P would also be decidable, which is a contradiction. Thus, the indirect argument uses a *reduction*: an algorithm for solving P' yields an algorithm for solving P. In this fashion we use problems already known to be undecidable to generate new undecidable problems. What has been said above can of course also be translated into the terminology of partial recursive functions.

As an illustration of the power of Church's Thesis as a proof method, we establish the existence of a *universal Turing machine*, that is, a specific Turing machine capable of simulating any other Turing machine. We return to our original definition of partial recursive functions $f(x_1, \ldots, x_k)$ of several variables. Thus, our Turing machines also compute functions of several variables, the input in this case being $(x_1, \ldots, x_k)$. Consider our enumeration $TM_0, TM_1, \ldots$ of Turing machines. If $f(x_1, \ldots, x_k)$ is the partial recursive function of k variables computed by the machine TM_i, we say that i is an *index* for f and write

$$f(x_1, \ldots, x_k) = f_i(x_1, \ldots, x_k).$$

Theorem 1.4. For every $k \geq 1$, there is a u such that

$$f_i(x_1, \ldots, x_k) = f_u(i, x_1, \ldots, x_k).$$

Proof. Consider the following intuitive algorithm for computing a function of $k+1$ variables. Given a $(k+1)$-tuple $(i, x_1, \ldots, x_k)$ as the input, find TM_i and start it with the k-tuple $(x_1, \ldots, x_k)$ as the input. If and when the computation halts, output the final result of the computation.

By Church's Thesis, the intuitive algorithm can be carried out by a Turing machine TM_u. Hence, u is an index for our (partial recursive) function of $k+1$ variables.
□

It is of course possible to give a detailed construction of TM_u. How this is done, is already hinted at in Turing's original article [Tur]. The machine TM_u is customarily referred to as the *universal Turing machine* (for partial recursive functions of k variables). Whenever TM_u receives as its input the index i of a particular Turing machine TM_i followed by some k-tuple $(x_1, \ldots, x_k)$, then TM_u behaves, as far as the output is concerned, exactly as TM_i would behave given the input $(x_1, \ldots, x_k)$.

1.2 Recursive enumerability, machine models and complexity

The recursiveness of a set S of nonnegative integers means that its characteristic function is computed by a Turing machine. By Church's Thesis, this is equivalent to saying that there is an algorithm for deciding whether or not a given integer belongs to S. A weaker requirement is to ask for an effective enumeration of the numbers in S. Sets satisfying this requirement are called *recursively enumerable*. We give the following formal definition.

Definition. A set of nonnegative integers is *recursively enumerable* iff it equals the domain of a partial recursive function $f(x)$.

Recursive enumerability

Thus, S being recursively enumerable means the existence of a Turing machine TM such that TM halts exactly in case its input x belongs to S. If x does not belong to S, we do not learn about this fact just by observing the computation of TM with the input x. No matter how long the computation has been going on without halting, we can never be sure that it will not halt at the next step!

The following result is immediate by the definitions.

Theorem 1.5. *Every recursive set is recursively enumerable.*

Proof. Assume that TM computes the characteristic function of a set S. Modify TM to a machine TM' by starting a loop from TM's output 0, and leaving TM unchanged otherwise. It follows that S equals the domain of the partial recursive function computed by TM', which shows that S is recursively enumerable.

□

Assume that S is recursively enumerable. Thus, for some Turing machine TM, S consists of all numbers x such that TM halts with x as its input. How does this give an effective enumeration of S? We can of course look at the computations of TM with different inputs and see what happens. But suppose the computation with input 2 loops, that is, runs forever. If we just look at the computations one after the other, we are never able to even start the computation with input 3.

This can be avoided by switching between different computations, never spending too much time with one computation. This technique, customarily referred to as *dovetailing*, is applicable to any sequence of computations C_0, C_1, C_2, ..., and consists of arranging the individual steps in different computations in such a linear order that each step in each computation eventually comes up. This linear order can be explicitly defined in terms of the *pairing function* discussed in the preceding section. Thus, the steps in the effective enumeration procedure are as follows. (Here we start the numbering from 0. The computation C_i in this case is $TM(i)$, the computation of TM with input i.)

$$
\begin{aligned}
0 &: \text{Step } 0 \text{ in } TM(0), \\
1 &: \text{Step } 0 \text{ in } TM(1), \\
2 &: \text{Step } 1 \text{ in } TM(0), \\
3 &: \text{Step } 0 \text{ in } TM(2), \\
4 &: \text{Step } 1 \text{ in } TM(1), \\
5 &: \text{Step } 2 \text{ in } TM(0), \\
6 &: \text{Step } 0 \text{ in } TM(3), \\
&\vdots
\end{aligned}
$$

In general, Step x in the effective enumeration procedure consists of Step $\varphi_1(x)$ in $TM(\varphi_2(x))$, where φ_1 and φ_2 are the inverse components of the pairing function. Whenever Step x is a halting one, the enumeration procedure produces the output $\varphi_2(x)$.

Whenever S is recursively enumerable, S can be effectively enumerated without repetition. Indeed, the procedure above enumerates S without repetition, provided it is understood that there are no steps in a computation after halting. A recursively enumerable set can also be enumerated in increasing order of its elements, but we have to be a bit cautious about what this actually means. After producing a new number to the list, we can always update the ordering of the numbers produced so far. However, if we are interested in, say, numbers less than 1000, we do not know in general when we have produced all of them.

The interconnection between recursive and recursively enumerable sets will become apparent in our next result.

Theorem 1.6. A set S is recursive iff both S and its complement $\sim S$ are recursively enumerable. There are recursively enumerable sets that are not recursive.

Proof. Consider the first sentence. Assume first that S is recursive. Then also $\sim S$ is recursive because its characteristic function is obtained from that of S by interchanging the outputs 0 and 1. Hence, by Theorem 1.5, both S and $\sim S$ are recursively enumerable. Assume, conversely, that the latter condition holds. Hence, there are effective procedures for listing both the elements $a_1, a_2, \ldots$ of S and the elements $b_1, b_2, \ldots$ of $\sim S$. Given an arbitrary integer x, we go through the two lists in a zig-zag fashion until we find x. This constitutes an algorithm for computing the characteristic function of S and, hence, S is recursive.

To prove the second sentence, we claim that the set

$$SA = \{x | TM_x \text{ halts with } x\},$$

consisting of the indices of all self-applicable Turing machines, is recursively enumerable but not recursive. Indeed, SA being recursive would contradict Theorem 1.3. That SA is recursively enumerable is seen by dovetailing through all computations $TM_x(x)$. □

The proof shows that the complement $\sim SA$ is not recursively enumerable because, otherwise, SA would be recursive. Thus, recursively enumerable sets are not *closed* under complementation, whereas recursive sets are closed under complementation: the complement of a recursive set is always recursive. It follows directly by the definitions that both recursively enumerable and recursive sets are closed under union and intersection. For instance, the characteristic function of $S_1 \cup S_2$ can be immediately computed from those of S_1 and S_2.

We have already pointed out in the preceding section that Turing machines can be modified in various ways without affecting the class of partial recursive functions. For instance, in point (ii) in our evidence given in support of Church's Thesis, several such modifications were listed. A further modification will be met in Chapter 5. A formal proof, often very tedious, can be given in regard to the equivalence of such modifications. Such a proof concerning the equivalence of models A and B consists of showing how every partial function computed by a specimen of model A can be computed by a specimen of model B, and vice versa. In other words, one shows how to *simulate* one of the models by the other.

Recursive enumerability

A great variety of machine models equivalent to Turing machines has been proposed. One such model, the *register machine*, will be useful for our purposes in Chapter 3. In this model the memory is structured differently than in Turing machines, being of random access rather than sequentially accessible. A register machine has finitely many separately accessible registers, each of them capable of holding an (arbitrarily large) integer. The registers $R_1, \ldots, R_r$ correspond to the tape of a Turing machine in that they serve as input, output and memory devices. The machine has finitely many internal states. To each state q_i there is associated a specific register, as well as a specific action. In q_i the specific action is always performed on the specific register. Only two actions are possible: add 1 to the contents of the register, or subtract 1 from the contents. The contents always remain nonnegative, so the latter action is performed only if the register contains a positive number. We say that the register is *empty* if it contains the number 0.

Assume that there are two registers R_1 and R_2, containing at some moment the numbers 630 and 47. Assume, further, that the machine is in state q_7, where the command "increment R_2 by 1" is associated. Then the computation step can be illustrated by the following diagram.

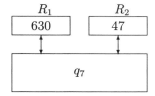 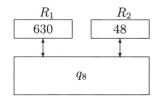

The numbers in the registers are given in unary representation: R_2 is a stack of 47 1s. The machine (that is, the box being in state q_i) has no idea about how many 1s a specific register contains except that it sees if there are no 1s. This information is vital for preventing a subtraction leading to a negative number. We have not yet indicated why q_8 is the "next state" in the illustration above.

The reason is that a register machine is defined by giving its *program*. A program is a sequence of *commands*, each provided with a *label* (or an address). The labels q_i are above called states, in analogy with Turing machines. The number of states being finite corresponds to the fact that the program consists of finitely many labelled commands. Each of the latter is of one of the following two forms:

$$q_i : R_j + 1 \text{ or } q_i : R_j - 1 \ (q_k),$$

where $k \neq i + 1$. More explicitly, the two commands are read as follows:

$$ADD \ 1 \ \text{to} \ R_j \ \text{and} \ GO \ TO \ q_{i+1}.$$

IF $R_j = 0$, *GO TO* q_k; *ELSE SUBTRACT* 1 from R_j and *GO TO* q_{i+1}.

In addition, the program contains a specific command *STOP* at the end of the list.

The program is read in the natural order. The only jumps are caused by the subtraction commands when the register is empty. Because of the jumps the computation may also run into a loop, and the *STOP* command is never reached.

There are various techniques for reducing the length of a program. For instance, by keeping one register always empty, we have plain *GO TO* commands available. The use of commands such as $R_j + 3$ and of several simultaneous additions can easily be reduced to the use of our basic commands at the cost of adding the number of labels. We do not need such techniques since we do not enter in any detail the constructions with register machines.

Example 1.4. The register machine defined below is the essential tool needed in the simulation of Turing machines by register machines. The machine has three registers R_1, R_2, R_3. (We use plain *GO TO* commands. We can avoid them by introducing a fourth register that is kept empty.) The labels are marked simply by i rather than by q_i. The input is given in register R_1, the two other registers being empty at the beginning. The output is read from R_1 and R_3, that is, we have in this case two outputs:

$$\begin{aligned}
1 &: R_1 - 1 \ (7), \\
2 &: R_3 + 1, \\
3 &: R_1 - 1 \ (7), \\
4 &: R_2 + 1, \\
5 &: R_3 - 1 \ (1), \\
6 &: GO\ TO\ 1, \\
7 &: R_2 - 1 \ (10), \\
8 &: R_1 + 1, \\
9 &: GO\ TO\ 7, \\
10 &: STOP.
\end{aligned}$$

For input 7, the machine goes through the computation steps described in the following table. The active command and the contents of each register are given for each time instant.

Recursive enumerability

Time	Command	R_1	R_2	R_3
0	1	7	0	0
1	2	6	0	0
2	3	6	0	1
3	4	5	0	1
4	5	5	1	1
5	6	5	1	0
6	1	5	1	0
7	2	4	1	0
8	3	4	1	1
9	4	3	1	1
10	5	3	2	1
11	6	3	2	0
12	1	3	2	0
13	2	2	2	0
14	3	2	2	1
15	4	1	2	1
16	5	1	3	1
17	6	1	3	0
18	1	1	3	0
19	2	0	3	0
20	3	0	3	1
21	7	0	3	1
22	8	0	2	1
23	9	1	2	1
24	7	1	2	1
25	8	1	1	1
26	9	2	1	1
27	7	2	1	1
28	8	2	0	1
29	9	3	0	1
30	7	3	0	1
31	10	3	0	1

During the time instants 0–20, input 7 is divided by 2, the quotient 3 is stored in R_2, and the remainder 1 in R_3. This part of the computation empties the register R_1. The remainder of the computation, in time instants 21–30, only serves the purpose of transferring the contents of R_2 back to R_1.

These remarks hold also in general. Assume that the input is x. (Thus, x 1s appear at the beginning in register R_1, the two other registers being empty.) The computation halts for any x (that is, the $STOP$ command 10 is reached). At the end registers R_1 and R_3 contain the quotient and remainder when x is divided by 2. In other words, R_1 contains the greatest integer $[\frac{x}{2}]$ not exceeding $\frac{x}{2}$, whereas R_3 is

0 or 1, depending on whether x is even or odd. Observe that R_3 switches between 0 and 1 and has its proper value when the final phase starting from command 7 is entered. If we let 7 be the $STOP$ command and omit commands 8–10, we obtain the same output except that, now, $[\frac{x}{2}]$ appears in register R_2. Observe, finally, that R_3 is never empty when 5 is entered. Thus, instead of going to (1), we could have any other transition without any effect on the output.

□

Register machines are flexible in regard to the input and output format. If we want to compute a function of k variables, we can store the input in the first k registers. Register machines can also be used to define sets of nonnegative integers. A register machine is said to *define* or *accept* a set S of nonnegative integers iff S consists of all integers x such that the machine halts (that is, reaches the $STOP$ command) after receiving x as the input. In other words, the machine halts if it receives an input belonging to S but loops if its input is a nonnegative integer not in S. A set S is *definable* or *acceptable* by a register machine iff there is a register machine such that S is the set defined by the machine.

Theorem 1.7. A set of nonnegative integers is recursively enumerable iff it is definable by a register machine.

Proof outline. Clearly, the computation of a register machine constitutes an effective procedure and hence, by Church's Thesis, it can be simulated by a Turing machine. Consequently, every set definable by a register machine is recursively enumerable.

Conversely, assume that S is recursively enumerable. Hence, by the definition, S equals the domain of a function $f(x)$ computed by a Turing machine TM. We want to prove that S equals the domain of a function computed by a register machine. We outline an argument showing how TM can be simulated by a register machine RM.

TM is defined by specifying finitely many instructions of the form

$$(*) \qquad (q, a) \text{ yields } (q', a', m),$$

where q and q' come from a finite state set, $a, a' \in \{0, 1\}$ and $m \in \{L, R\}$. The instructions tell how an instantaneous description is changed into the next one. We show how the same change can be affected by RM.

We view instantaneous descriptions as triples

$$(**) \qquad (n_l, (q, a), n_r),$$

where TM scans a in state q, n_l are the contents of the tape to the left of the scanned position, and n_r are the contents of the tape to the right of the scanned position. Moreover, n_l and n_r are viewed as integers in binary notation in such a way that n_r is read from *right to left*. Thus, the instantaneous description

$$\ldots 10011 q_2 00101011 \ldots,$$

where the dots indicate infinite sequences of 0s, is, for our purposes, viewed as the triple
$$(19, (q_2, 0), 106),$$
where we have written n_l and n_r in decimal notation. More specifically, the significant portion of the tape to the right of the scanned 0 is 0101011. From right to left this reads 1101010, which is the binary representation of 106.

The numbers n_l and n_r are kept in two specific registers R_l and R_r, whereas the pairs (q, a) are some of the labels of the commands in the program of RM. Taking into account the instruction $(*)$, we are able to define the command with the label (q, a) in such a way that a computation C transforming $(**)$ in agreement with TM is initiated. The computation agrees with TM for all n_l and n_r. It uses some auxiliary registers and commands. (This is why we also need labels other than those of the form (q, a).) The computation C ends with the $GO\ TO$ command to the proper label (q', b).

The technique presented in Example 1.4 is the essential tool needed to define the command (q, a) in such a way that the proper computation C results. Let us consider this in more detail. We want to change the contents of the registers R_l and R_r to n'_l and n'_r and find a label (q', b) in such a way that

$(**)'$ $\qquad\qquad\qquad (n'_l, (q', b), n'_r)$

is the instantaneous description following $(**)$ in the computation of TM.

In $(**)'$, q' will be the same as in $(*)$. The other items in $(**)'$ are defined differently for the cases $m = L$ and $m = R$. Assume first that in $(*)$ we have $m = L$. Then
$$n'_l = [n_l/2], \quad b = \text{rem}(n_l/2), \quad n'_r = 2n_r + a',$$
where $\text{rem}(n_l/2)$ is the remainder of the division of n_l by 2. Indeed, the tape of TM to the right from the new position scanned consists of a' followed by everything that was there previously. This leads to the formula $n'_r = 2n_r + a'$. The other equations are obtained by a similar argument.

Let us return to the example considered earlier in this proof. Assume that we have the instruction
$$(q_2, 0) \text{ yields } (q_5, 1, L).$$
Then $1001q_5110101011$ is the instantaneous description following the instantaneous description $10011q_200101011$ in the computation of TM. The new instantaneous description is written in our notation of triples $(9, (q_5, 1), 213)$. We observe that
$$9 = [19/2] = [n_l/2], \quad 1 = \text{rem}(19/2), \quad 213 = 2 \cdot 106 + 1 = 2n_r + a'.$$

Similarly, if $m = R$, we obtain
$$n'_l = 2n_l + a', \quad b = \text{rem}(n_r/2), \quad n'_r = [n_r/2].$$

Altogether, we have to be able to replace the contents x in a register by one of the following: $[x/2], \text{rem}(x/2), 2x, 2x + 1$. We saw in Example 1.4 how to carry out the first two replacements. It is an easy exercise to carry out the latter two replacements. Example 1.4 also presented a technique for transferring a number from one register to another. Thus, we can always define a subprogram, starting with the command with the label (q, a) and ending with the appropriate *GO TO* command (to (q', b)), such that the contents of registers R_l and R_r are changed in the correct fashion.

□

It is by no means essential in the above simulation argument that the tape alphabet of the Turing machine consists of two letters. If there are k letters, we consider the k-ary notation for integers. (This becomes the ordinary decimal notation for $k = 10$.) The division by 2 now becomes division by k, but registers R_l and R_r are still stacks of 1s, that is, the notation in the registers is *unary*. (This is the reason why registers are often called *counters*.)

We will need register machines in Chapter 3. The model considered and explained in detail in Chapter 3 is a slight variant of the model introduced above. In particular, the subtraction command is divided into two consecutive commands: the test command including the jump instruction, and the actual subtraction command. Moreover, the registers are empty when the *STOP* command is reached. Such irrelevant changes are made for notational convenience. Observe that we can view our original *STOP* command as a "temporary" one, initializing a computation, where the registers are emptied, and ending with the "real" *STOP* command.

We conclude this section with a brief discussion about *computational complexity*. We have already pointed out the difference between effective and efficient procedures. Efficiency is of crucial importance in practical considerations. An algorithm for a problem is of little practical use if instances of a reasonable size take millions of years of computing time. For instance, in cryptography everything is decidable — it is the efficiency that matters. On the other hand, complexity issues are met with infrequently in the remainder of the book, notably in Section 4.3.

The *time complexity* of an algorithm is a function of the length of the input. An algorithm is of time complexity $f(n)$ iff, for all n and all inputs of length n, the execution of the algorithm takes at most $f(n)$ steps. If n is an integer, its length is the number of digits or bits in n. Thus, here we use binary or decimal notation. For unary notation, where the length of n equals n itself, some of the statements given below become untrue.

Clearly, the time complexity depends on the *machine model* we have in mind. The number of steps becomes smaller if more work can be included in one step. Some algorithms run faster on a register machine than on a Turing machine, and vice versa. However, fundamental notions such as polynomial time complexity are independent of the model. This concerns only models chosen with "good taste". One should not include in one step an abstract subroutine for a complex problem, such as the testing of the primality of a given integer.

There are often slow and fast algorithms for the same problem. In some cases even an unlimited speed-up is possible. In general, it is difficult to establish *lower bounds* for complexity, that is, to show that every algorithm for a specific problem is of at least a certain time complexity, say possesses a quadratic lower bound. The considerations below deal with *upper bounds*.

Natural complexity measures are obtained by considering Turing machines. How many steps (in terms of the length of the input) does a computation require? This is a natural formalization of the notion of time resource. Similarly, the number of squares visited during a computation constitutes a natural space measure. As pointed out above, one has to be careful in defining the length of the input. One can also change the input format. To obtain lower than linear space complexities, the input has to be given on a separate tape, and only the squares of the work tape are counted.

Let us now be more specific. Consider a Turing machine TM that halts for all inputs. This assumption is reasonable, because we are considering decidable problems or recursive sets, and are interested in subclasses possessing efficient algorithms. The *time complexity function* associated with TM is defined by

$$f_{TM}(n) = \max\{m | TM \text{ halts after } m \text{ steps for an input of length } n\}.$$

The Turing machine TM is *polynomially bounded* iff there is a polynomial $p(n)$ such that
$$f_{TM}(n) \leq p(n)$$
holds for all n. The notation **P** is used for all problems that can be solved by a polynomially bounded Turing machine. (Using a suitable encoding, we can view any problem as a problem of computing the characteristic function of a set of nonnegative integers.)

A problem is referred to as (*computationally*) *intractable* (sometimes also *computationally impossible*) if it is not in **P**. Tractable problems have several natural subclasses: problems with linear, quadratic, cubic time complexity. The informal reference to a problem as *easy* means that the values of the polynomial are small, at least within the range considered. Because of the difficulty in establishing lower bounds, intractability means in most cases that no good algorithm is *known at present*. One-way functions $f(x)$ are important in cryptography and also in Section 4.3: given the definition of f, it is easy to compute $f(x)$ from x, whereas the computation of x from $f(x)$ is likely to be intractable.

A *nondeterministic Turing machine* NTM may have several possibilities for its behaviour, that is, several possible instructions for a pair (q, a). Thus, an input gives rise to several computations. This can be visualized as the machine making guesses or using an arbitrary number of parallel processors. For an input w, we denote by $s(w)$ the shortest halting computation starting from w. The time complexity function is now defined by

$$f_{NTM}(n) = \max\{1, m | w \text{ is of length } n \text{ and } s(w) \text{ has } m \text{ steps}\}.$$

The pair $(1, m)$ is considered because possibly no input of length n for our nondeterministic Turing machine leads to a halting computation, for some values of

n. (We emphasize again that our notation here is not unary. In unary notation, there is only one input of each length.) The notions of a polynomially bounded nondeterministic Turing machine and the corresponding class of problems, **NP**, are now defined exactly as in the deterministic case.

Problems in **P** are tractable, whereas problems in **NP** have the property that it is tractable to check whether or not a guess for the solution is correct. It is not known whether the factorization of an integer is in **P** but it certainly is in **NP**: one guesses the decomposition and verifies the guess by computing the product.

By definition, **P** is included in **NP** but it is a celebrated open problem whether or not **P** = **NP**. However, there are many **NP**-complete problems. A specific problem is **NP**-*complete* iff it is in **NP** and, moreover, it is **NP**-*hard*, that is, every problem in **NP** can be reduced in polynomial time to this specific problem. (A formal definition for this reduction is in terms of Turing machines: for an arbitrary problem in **NP**, there is a polynomially bounded deterministic Turing machine that translates every instance of the arbitrary problem into an instance of our specific problem.) It follows that **P** = **NP** iff an **NP**-complete problem is in **P**. In such a case an arbitary problem in **NP** can be settled in deterministic polynomial time because it can be reduced in polynomial time to the specific **NP**-complete problem which, in turn, can be settled in polynomial time. It is generally conjectured that **P** $\neq$ **NP**. Therefore, **NP**-complete problems are considered to be intractable.

A specific problem in **NP** can be shown to be **NP**-complete by proving that some problem previously known to be **NP**-complete can be reduced in polynomial time to the specific problem in question. However, we need something to start with: a problem whose **NP**-completeness can be established by a direct argument. A problem very suitable for this purpose is the *satisfiability problem for well-formed formulas of the propositional calculus*: does a given formula assume the truth-value "true" for some assignment of truth-values for the variables. The formulas are built from variables using propositional connectives (conjunction, disjunction, negation). The satisfiability problem is basic because it is capable of encoding Turing machine computations: the computation of a given Turing machine with a given input being successful is equivalent to a certain well-formed formula being satisfiable.

Although the classes **P** and **NP** remain invariant in the transition from one machine model to another, the degree and coefficients of the polynomial involved may change.

Space complexity and the classes **P** $-$ $SPACE$ and **NP** $-$ $SPACE$ are defined analogously. Clearly, a time class is included in the corresponding space class because one time unit is needed to extend the space by one square. However, it is not difficult to prove that **P** $-$ $SPACE$ = **NP** $-$ $SPACE$. Consequently, we have the following chain of inclusions:

$$\mathbf{P} \subseteq \mathbf{NP} \subseteq \mathbf{NP} - SPACE = \mathbf{P} - SPACE \,.$$

It is a celebrated open problem whether or not the two inclusions are proper.

1.3 Language theory and word problems

The notion of a (formal) language introduced below is very general. It certainly includes both natural and programming languages. In this section we consider some notions and tools needed in later chapters. In addition, we show how the concept of a *rewriting system* can be used to formalize various models of computation. In Section 1.1 we did not present any details of a formal definition about Turing machine computations. The details are obtained as a by-product of the discussion concerning rewriting systems.

We begin with some very fundamental language-theoretic notions. An *alphabet* is a finite nonempty set, often denoted by Σ of V. Its elements are referred to as *letters*. *Words* are finite strings of letters. The *empty word* λ is viewed as the string consisting of zero letters. The (infinite) set of all words over Σ is denoted by Σ^*. Similarly, Σ^+ is the set of all nonempty words.

The juxtaposition uv of two words u and v is called their *catenation*. The empty word acts as an identity: $w\lambda = \lambda w = w$ holds for all words w. The notation w^i means catenating i copies of w, with $w^0 = \lambda$. The *length* $|w|$ of a word w is the number of letters in w, each letter being counted as many times as it occurs. Thus,

$$|uv| = |u| + |v| \text{ and } |w^i| = i|w|,$$

for all words u, v, w and integers $i \geq 0$. We also use vertical bars to denote the absolute value of a number. No confusion will arise because the meaning will be clear from the context.

A word u is a *subword* (or a *factor*) of a word w iff $w = xuy$ holds for some words x and y. If $x = \lambda$ (resp. $y = \lambda$) here then u is a *prefix* (resp. *suffix*) of w.

Some facts about words are rather obvious. Assume that $uv = xy$. Then x is a prefix of u, or vice versa. If the former alternative holds, there is a unique word z such that $xz = u$ and $zv = y$. The following result is often referred to as the *Lyndon Theorem*.

Theorem 1.8. Assume that $uv = vw$ holds for some words u, v, w. Then there are words x, y and an integer $i \geq 0$ such that

(*) $$u = xy, \ w = yx, \ v = (xy)^i x = x(yx)^i.$$

Consequently, whenever two words u and v commute, $uv = vu$, they must be powers of the same word.

Proof. We apply induction on the length of v. If $|v| = \lambda$, then $u = w$. We choose $x = \lambda$, $y = u$ and $i = 0$ to satisfy (*). Assume that the claim (*) holds for $|v| < k$ and consider the equation

$$uv = vw, \ |v| = k.$$

If $|v| \leq |u|$, we may write

$$u = vz, \ w = zv,$$

for some z. Again the conclusion $(*)$ holds, now with $x = v$, $y = z$ and $i = 0$. If $|u| < |v|$, we write $v = ut$, whence $u^2 t = utw$. Consequently,
$$ut = tw \ , \ \ |t| < |v| \ .$$
(The inequality follows by $v = ut$, because we may exclude the trivial case $u = \lambda$.) By the inductive hypothesis, we may write
$$u = x_1 y_1 \ , \ \ w = y_1 x_1 \ , \ \ t = (x_1 y_1)^j x_1 = x_1 (y_1 x_1)^j \ ,$$
for some x_1, y_1 and j. Because $v = ut$, $(*)$ now follows for the choice $x = x_1$, $y = y_1$, $i = j + 1$.

To prove the last sentence, assume the contrary: $uv = vu$ but u and v are not powers of the same word. Assume also that in this counterexample $|uv|$ assumes its minimal value, that is, there are no shorter counterexamples. We apply the first part of the theorem for the case $u = w$. Thus, for some x and y, $xy = yx = u$. Because $|xy| < |uv|$, x and y are powers of the same word. This implies, by $(*)$, that u and v are powers of the same word, a contradiction. □

Any subset L of Σ^* is referred to as a *(formal) language* over Σ. A finite language can, at least in principle, be defined by listing all of its words. A finitary specification is needed for the definition of an infinite language. General specification methods will be given below.

Since languages are sets, we may apply *Boolean operations* to languages and speak of the *union* $L \cup L'$, *intersection* $L \cap L'$ and *difference* $L - L'$ of two languages L and L', as well as of the *complement* $\sim L$ of a language L. The *catenation* of two languages is defined by
$$L_1 L_2 = \{w_1 w_2 | w_i \in L_i \ , \ \ i = 1, 2\} \ .$$
Power notation is used with the convention $L^0 = \{\lambda\}$. The operations of *catenation closure* (also called *Kleene star* and *Kleene plus*) are defined by
$$L^* = \bigcup_{i=0}^{\infty} L^i \ \ \text{and} \ \ L^+ = \bigcup_{i=1}^{\infty} L^i \ .$$
Observe that the notations Σ^* and Σ^+ are in accordance with this definition.

An operation often met with in the rest of the book is the operation of *morphism*. A mapping $h : \Sigma^* \to \Delta^*$, where Σ and Δ are alphabets, satisfying the condition
$$h(uv) = h(u)h(v) \ , \ \ \text{for all} \ u, v \in \Sigma^* \ ,$$
is called a *morphism*. A morphism h is *nonerasing* iff $h(a) \neq \lambda$, for all $a \in \Sigma$. (In algebraic terminology, Σ^* is a free *monoid* with identity λ, and the morphism above is a monoid morphism.) The definition is extended to languages:
$$h(L) = \{h(w) | w \in L\} \ .$$

Language theory and word problems 27

Many issues concerning decidability and undecidability, important in our later considerations, can be presented and illustrated using language-theoretic terms. Arithmetical operations are inferior to language-theoretic ones for many purposes. This will become apparent in particular in Chapters 2 and 4. Because language theory is very useful in the presentation of many issues that are important for us, we also have to present some facts that are not directly connected with decidability theory.

However, we want to emphasize that we can always replace languages by sets of numbers, and vice versa. The transition procedure is effective, even efficient. We now describe a natural way to perform the transition.

Consider words over the alphabet $\{a, b\}$. If we view the letters as bits, $a = 0$, $b = 1$, every word becomes a number in *binary* notation. For instance, $baaab = 10001 = 17$. Also, every word $a^i baaab$, $i \geq 0$, becomes the number 17. This means that this correspondence between words and numbers is not one-to-one. A one-to-one correspondence results if *dyadic* rather than binary notation is used. This means that 2 continues to be the base of the number system but the digits are 1 and 2: $a = 2$ and $b = 1$ (or vice versa). Now

$$baaab = 12221 = 1 \cdot 2^4 + 2 \cdot 2^3 + 2 \cdot 2^2 + 2 \cdot 2 + 1 = 45 \ .$$

If we agree that the empty word λ corresponds to 0, we now have a one-to-one correspondence between Σ^* and the set of nonnegative integers.

Words over an alphabet with n letters are viewed similarly to n-*adic* integers: the corresponding number system has the base n and digits $1, \ldots, n$. The letters of the alphabet have to be ordered, in order to obtain a correspondence between the letters and the digits. For instance, define for the alphabet $\{a, b, c\}$ the correspondence $a = 1$, $b = 2$, $c = 3$. Then the word $abca$ becomes the number

$$1 \cdot 3^3 + 2 \cdot 3^2 + 3 \cdot 3 + 1 = 55 \ .$$

No other word is mapped to 55.

In general, for an alphabet Σ with n letters, the n-adic correspondence between Σ^* and the set of nonnegative integers is one-to-one. This correspondence can be extended to cover arbitrary languages over a specific Σ and arbitrary sets of nonnegative integers. It is important to notice that a different correspondence results if the cardinality of Σ is changed. If one begins with a set of numbers, one has to fix the size of the alphabet in order to obtain a unique language. (Of course, renaming the letters does not effect any essential change.) A change in the cardinality of the alphabet may result in a huge increase in the descriptional complexity of the language.

Example 1.5. Consider the set

$$S = \{2^n - 1 | n \geq 1\} \ .$$

If we choose the binary alphabet $\{1, 2\}$, the corresponding language has a very simple description

$$1^+ = \{1^i | i \geq 1\} \ .$$

On the other hand, using the choice $\{1,2,3\}$ for the alphabet we obtain a language corresponding to S which is rather difficult to describe. The first few words, giving the 3-adic representation of the numbers 1, 3, 7, 15, 31, 63, 127 are:

$$1,\ 3,\ 21,\ 113,\ 311,\ 1323,\ 11131.$$

Next, assume that S consists of all numbers incongruent to 2 modulo 3. Now it is natural to choose the alphabet $\{1,2,3\}$. The language $L(S)$ corresponding to S consists of all words except those ending with the letter 2. Using the operations discussed above, we obtain

$$L(S) =\sim (\{1,2,3\}^*2).$$

Other choices of the alphabet lead to much more complicated languages. However, we can obtain an amazingly simple "quasi-dyadic" description for S. If we use the digits 1 and 4, rather than 1 and 2, it can be shown inductively that

$$S = \{1,4\}^*.$$

Indeed, the words λ, 1, 11, 4, 14, 111, 41, 114, 44 represent the numbers ≤ 12 in S, in increasing order. Observe that the number 3, essential in the definition of S, is not directly involved in this representation.

As a final illustration, assume that S consists of all nonnegative integers *not* of the form $2^i - 3$, $i \geq 2$. Thus, 1, 5, 13, 29, 61, are the first few missed numbers. In dyadic notation the missed numbers are $2^i 1$, $i \geq 0$. Thus, $\sim (2^*1)$ is the language corresponding to S. We also obtain a "direct" representation for S, not using complementation at all. The idea is similar to the one used in the preceding illustration. Let us still use 2 as the base of a number system but consider $2, 3, 4$ as digits. Thus, 23343 represents the number

$$2 \cdot 2^4 + 3 \cdot 2^3 + 3 \cdot 2^2 + 4 \cdot 2 + 3 = 79.$$

It can be shown that $S = \{2,3,4\}^*$.

□

We now introduce the basic definitional tool in language theory. A *rewriting system* (often called also *semi-Thue system*) is a pair

$$RW = (\Sigma, P),$$

where Σ is an alphabet and P is a finite set of ordered pairs of words over Σ. The elements (u,v) of P are referred to as *rewriting rules* or *productions* and are written $u \to v$.

The idea in the rewriting process is that an occurrence of u as a subword can be rewritten as v. Formally, we define the *yield relation* $\Rightarrow_{RW}$ (or briefly $\Rightarrow$ if RW is understood) as follows. For any words x and y in Σ^*, $x \Rightarrow y$ holds iff

$$x = z_1 u z_2 \text{ and } y = z_1 v z_2,$$

Language theory and word problems 29

for some words z_1 and z_2 and some production $u \to v$ in P. We say that x *yields y directly according* to RW.

The word x *yields* the word y, in symbols $x \Rightarrow^* y$, iff there is a finite sequence of words
$$x = z_0, z_1, \ldots, z_k = y \text{ with } z_i \Rightarrow z_{i+1}, \ 0 \le i \le k-1.$$
We also use the notation
$$z_0 \Rightarrow z_1 \Rightarrow \ldots \Rightarrow z_k$$
and refer to the sequence as a *derivation*.

A *Thue system* has bidirectional productions $u \leftrightarrow v$. This means that u may be replaced by v and v may be replaced by u. A Thue system can always be viewed as a semi-Thue system, where the productions are given as pairs $u \to v$ and $v \to u$. We will also consider the difference between Thue and semi-Thue systems in connection with word problems and in Chapter 4.

A rewriting system can be converted into a language-defining device in various ways. One may choose a finite set of *start words* (or *axioms*) and consider the language consisting of all words derivable from some of the axioms. Similarly, one may choose a finite set of *target words* and consider the language consisting of all words deriving one of the target words. The customary notion of a (*phrase-structure*) *grammar* belongs to the former type.

A *grammar* is a quadruple $G = (RW, N, T, S)$, where $RW = (\Sigma, P)$ is a rewriting system, N and T are disjoint alphabets with $\Sigma = N \cup T$, and S is in N. The *language generated* by G is defined by
$$L(G) = \{w \in T^* | S \Rightarrow^*_{RW} w\}.$$

Two grammars are *equivalent* iff they generate the same language.

Thus, not everything produced in the rewriting process belongs to the generated language. A *filtering device* is applied to the set of words derived from the *start letter S* (called *sentential forms*). Only sentential forms over the *terminal* alphabet T go through the filter. The *nonterminal* letters in N are auxiliary and serve the purpose of excluding intermediate words from the language. We often specify a grammar simply by listing the productions. We then apply the *convention* that nonterminals are capital letters, whereas terminals are lower case letters.

Example 1.6. For the grammar G with the two productions $S \to aSb$ and $S \to ab$, we have
$$L(G) = \{a^i b^i | i \ge 1\}.$$
Consider next the grammar G with the productions
$$S \to aSb, \ S \to bSa, \ s \to SS, \ s \to \lambda.$$
Denote by $\#_a(w)$ the number of occurrences of the letter a in w, similarly $\#_b(w)$. Then
$$L(G) = \{w | \#_a(w) = \#_b(w)\}.$$

Indeed, it can be seen immediately from the productions that every word in $L(G)$ has equally many as and bs. Conversely, all words with this property are in $L(G)$. This can be shown inductively, using the following fact. Every nonempty word w with equally many as and bs is of one of the forms aw_1b, bw_1a or w_2w_3, where each w_i has equally many as and bs, and w_2 and w_3 are nonempty.

Consider, finally, the grammar G determined by the productions

$$S \to abc,\ S \to aAbc,\ Ab \to bA,\ Ac \to Bbcc,$$

$$bB \to Bb,\ aB \to aaA,\ aB \to aa\ .$$

We claim that
$$L(G) = \{a^n b^n c^n | n \geq 1\}\ .$$

In fact, every derivation according to G begins with an application of the first or second production, the first one leading directly to termination. Arguing inductively, we assume that we have produced the word $a^i A b^i c^i$, $i \geq 1$. The only possibility of continuing the derivation is to let A travel to the right (production $Ab \to bA$), deposit bc when c^i is reached ($Ac \to Bbcc$) and let B travel to the left ($bB \to Bb$). We now have the sentential form $a^i B b^{i+1} c^{i+1}$, from which one of the two words

$$a^{i+1} A b^{i+1} c^{i+1} \quad \text{or} \quad a^{i+1} b^{i+1} c^{i+1}$$

is obtained by the last two productions. Every step in the derivation is uniquely determined except the last one, where we either terminate or enter a new cycle. □

We now give a formalization of Turing machines in terms of rewriting systems. In this formalization Turing machines accept languages rather than sets of numbers. However, this difference is unessential with regard to decidability. The formalization in terms of rewriting systems is not very intuitive but it is the shortest in terms of the formal apparatus developed so far.

Definition. A rewriting system $TM = (\Sigma, P)$ is called a *Turing machine* iff conditions (i)–(iii) are satisfied.

(i) Σ is divided into two disjoint alphabets Q and T, referred to as the *state* and the *tape* alphabets.

(ii) Elements q_0 and q_F of Q, the *initial* and the *final* state, as well as elements $\#$ and B of T, the *boundary marker* and the *blank symbol*, are specified. Also a subset T_1, the *terminal* alphabet, of the remaining elements of T is specified. It is assumed that T_1 is not empty.

(iii) The productions of P are of the forms

(1) $$q_i a \to q_j b \quad \text{(overprint)},$$

(2) $$q_i ac \to aq_i c \quad \text{(move right)},$$

(3) $\qquad q_i a\# \to a q_i B\#$ (move right and extend workspace),

(4) $\qquad\qquad c q_i a \to q_i c a$ (move left),

(5) $\qquad \# q_i a \to \# q_i B a$ (move left and extend workspace),

where the qs are in Q, $q_i \neq q_F$, and a, b, c are in T but different from $\#$. For each pair (q_i, a), where q_i and a are in their appropriate ranges, P either contains no productions (2) and (3) (resp. (4) and (5)) or else contains both (3) and (2) for every c (resp. contains both (5) and (4) for every c). There is no pair (q_i, a) such that the word $q_i a$ is a subword of the left side in two productions of the forms (1), (3), (5).

The Turing machine TM *halts* with a word w iff there are words w_1 and w_2 such that

$$\# q_0 w \# \Rightarrow^*_{TM} \# w_1 q_F w_2 \# .$$

Otherwise, TM *loops* with w. The language $L(TM)$ *accepted* by TM consists of all words over the terminal alphabet T_1 for which TM halts. □

One rewriting step according to the definition given above corresponds to the transition from an instantaneous description to the next one, as explained in Section 1.1. A slight difference with respect to Section 1.1 is that the above TM either only overprints or moves the read–write head during one time instant, rather than doing both of the operations at the same time instant. In other words, instructions can be viewed as *quadruples* instead of *quintuples*. This is no essential difference because, using an auxiliary state, the two operations can be performed one after another.

The use of the boundary marker $\#$ formalizes the idea that at any moment only a finite portion of the tape contains nonblank symbols. Productions (3) and (5) guarantee that this portion is potentially infinite. Because of the further conditions imposed on the productions, TM is *deterministic* and its behaviour depends only on the present state and the scanned letter.

We say that a language L is *recursively enumerable* iff L is acceptable by some Turing machine, that is, $L = L(TM)$ for some TM. The *characteristic function* $f_L : \Sigma^* \to \{0, 1\}$ of a language $L \subseteq \Sigma^*$ is defined in the following natural fashion:

$$f_L(w) = \begin{cases} 1 & \text{if } w \in L, \\ 0 & \text{otherwise}. \end{cases}$$

A language L is *recursive* iff its characteristic function is recursive.

It is clear that there is no essential difference with respect to the notions of recursiveness and recursive enumerability of sets, introduced before. In particular, a language is recursive (resp. recursively enumerable) iff the corresponding set of numbers is recursive (resp. recursively enumerable). The correspondence is defined in the way discussed above. Here the choice of alphabet is irrelevant in the transition from a set of numbers to the corresponding language. Theorem 1.6 now assumes the form.

Theorem 1.9. A language L is recursive iff both L and $\sim L$ are recursively enumerable. There are recursively enumerable languages that are not recursive.

Word problems constitute a very natural, as well as an old and widely studied, class of decision problems. We begin with finitely many equations

$$w_1 = w'_1, \ldots, w_k = w'_k,$$

where the ws are words. We want to find an algorithm for deciding whether or not an arbitrary given word w can be transformed into another given word w' using the equations. Without any formal definition, the meaning should be clear: we are allowed to substitute a subword w_i by w'_i, or vice versa, and to do this as many times as we like.

The problem formulated above is the *word problem for semigroups*. (The equations define a particular finitely generated semigroup.) The *word problem for groups* is defined similarly. It is a special case of the word problem for semigroups because, in the case of groups, among the equations there are always particular ones stating the properties of the inverses and the identity.

Word problems are also referred to as *Thue problems*. The terminology has already been discussed above. The difference between Thue problems and *semi-Thue problems* arises from the concept of rewriting: bidirectional or unidirectional.

Theorem 1.10. The Thue problem for semigroups is undecidable. More specifically, there is a semigroup G with finitely many defining relations $w_i = w'_i$, $i = 1, \ldots, k$, such that there is no algorithm for deciding whether or not an arbitrary equation $w = w'$ holds in G.

Proof. In our definition of a Turing machine, the final state q_F appears in no production on the left side. We now modify the definition by adding all productions

$$aq_F \to q_F \quad \text{and} \quad q_F a \to q_F,$$

where a ranges over the tape alphabet. The effect of these productions is that all letters other than q_F are erased. Consequently, a Turing machine halts with a word w iff

$$(*) \qquad \qquad \#q_0 w\# \Rightarrow^*_{TM} q_F.$$

It is clear that this change does not affect the language accepted by TM.

We now apply Theorem 1.9 and choose a Turing machine TM, modified in the way indicated above, such that $L(TM)$ is not recursive. Consequently, there is no algorithm for deciding whether or not an arbitrary word w satisfies $(*)$.

Let G be the semigroup whose defining relations are obtained by transforming the productions of TM into equations. Thus, $w_i = w'_i$ is a defining relation of G iff $w_i \to w'_i$ is a production of TM. We now claim that $(*)$ holds iff

$(**)$ $$\#q_0 w\# = q_F$$

is valid in G. Clearly, Theorem 1.10 follows from this claim. Indeed, the claim implies that the word problem of G is undecidable even for words of special type appearing in $(**)$.

It is clear that whenever $(*)$ holds, then $(**)$ also holds. Assume, conversely, that $(**)$ holds. Consequently, there is a finite sequence of words

$$x_0 = \#q_0 w\# \ , \ x_1, \ldots, x_n = q_F$$

such that every x_i, $1 \leq i \leq n$, is obtained from x_{i-1} by using some of the defining relations of G. We have to show how we can handle the situation, where the substitution has been carried out in the wrong direction with regard to the productions of TM.

We may assume without loss of generality that all words x_i in the sequence are distinct because, otherwise, we may replace the original sequence by a shorter one. Let m be the smallest index such that q_F appears in x_m. We replace the final part $x_m, \ldots, x_n$ in the original sequence by a sequence obtained by applying only productions $aq_F \to q_F$ and $q_F a \to q_F$. Hence, the final part can be modified to conform with TM, and we only have to show that

$(*)'$ $$x_0 \Rightarrow^*_{TM} x_m \ .$$

We know that none of the words $x_0, \ldots, x_{m-1}$ contains an occurrence of q_F. If every word in this sequence results from the preceding one by replacing the left side of some production of TM by its right side, we conclude that $(*)'$ holds. Otherwise, let r be the greatest index such that x_r results from x_{r-1} by replacing the right side v of some production $u \to v$ of TM by its left side u. Clearly, $r \leq m-1$ because q_F is introduced for the first time at the step from x_{m-1} to x_m and, hence, this step uses a production in the correct direction. Thus, for some y and z,

$$x_{r-1} = yvz \quad \text{and} \quad x_r = yuz \ .$$

By the choice of r, x_{r+1} results from x_r by applying some production of TM. But $u \to v$ is the only production applicable to x_r. This follows because the state symbol determines the position of the rewriting, and there is only one possible behaviour because TM is deterministic. Hence, we reverse the preceding step and, consequently, $x_{r-1} = x_{r+1}$. This contradicts the assumption of the xs being distinct and, thus, we conclude that $(*)'$ holds. □

The proof does not list the defining relations of the semigroup G explicitly. We mention (see [Sal 2] for further details) that the semigroup with the defining relations

$$ac = ca \ , \ ad = da \ , \ bc = cb \ , \ bd = db \ ,$$

$$aba = abae = eaba \ , \ eca = ae \ , \ edb = be$$

has an undecidable word problem.

In the remainder of this section we give an overview of some language-theoretic concepts and results needed occasionally in the rest of the book. The overview is somewhat telegraphic. It is recommended that the reader consults this part only if the need arises.

The *Chomsky hierarchy* of grammars and languages is obtained by imposing restrictions on the form of the productions. The four resulting classes of grammars and languages are defined as follows.

Type 0. No restrictions on the form.

Type 1 or **context-sensitive.** All productions $u \to v$ are *length-increasing*, $|u| \leq |v|$. (The production $S \to \lambda$ is also allowed if S does not occur on the right side of any production.)

Type 2 or **context-free.** The left side of every production consists of a single nonterminal.

Type 3 or **regular.** The productions are of the two forms $A \to wB$ and $A \to w$, where A and B are nonterminals and w is a word over the terminal alphabet.

Theorem 1.11. The family of type 0 languages equals the family of recursively enumerable languages. For $i = 1, 2, 3$, the family of type i languages is strictly included in the family of type $i - 1$ languages.

For instance, the first language in Example 1.6 is context-free but not regular, whereas the last language is context-sensitive but not context-free.

Regular languages coincide with languages accepted by *finite automata*. A *finite automaton* is a read-only Turing machine whose head scans the input word from left to right. At each step the next internal state is determined by the preceding one and the scanned letter. At the beginning the automaton is in the initial state and scans the first letter. The input is accepted iff the automaton enters a final state (there may be several of them) after scanning the last letter. For a *nondeterministic* finite automaton an equivalent *deterministic* one can always be constructed. Regular languages also coincide with languages obtained from the empty language and singleton languages $\{a\}$, where a is a letter, by consecutive applications of *regular operations*: Boolean operations, catenation and Kleene star. The formula stating how the operations are used is referred to as a *regular expression*. Some regular expressions have already been presented in Example 1.5.

Context-free languages coincide with languages accepted by *pushdown automata*. A *pushdown automaton* is a finite automaton combined with a potentially infinite *pushdown stack*. Information is retrieved from the latter using the principle "first in–last out" (or "last in–first out"). At each computation step, the behaviour is determined by the internal state, the letter scanned from the input tape and the letter topmost in the stack. The behaviour consists of changing the state, moving the input tape one square to the left or keeping it fixed, and replacing the topmost letter in the stack by a word. The latter can also be the empty word, which means "popping up" the contents of the stack. The behaviour is *nondeterministic*: for each configuration, there may be several possibilities. Acceptance is as for finite automata, with the following additions. Initially the stack contains a specific start letter; its contents at the end are irrelevant. As always with nondeterministic models of computation, an input is accepted if it gives rise to one successful computation, no matter how many failures it causes.

Contrary to Turing machines and finite automata, *deterministic pushdown automata* accept a strictly smaller family of languages than do nondeterministic ones. Languages in this subfamily are referred to as *deterministic context-free* languages. We mention another family, the family of *unambiguous languages*, lying strictly between deterministic context-free and context-free languages. A derivation according to a context-free grammar is *leftmost* iff at every step the leftmost nonterminal is rewritten. A context-free grammar is *unambiguous* iff no word in its language possesses two distinct leftmost derivations. A context-free language is *unambiguous* iff it is generated by some unambiguous grammar.

A context-free grammar is *linear* iff the right side of every production contains at most one nonterminal. Regular grammars are a special class of linear grammars. *Linear languages*, that is, languages generated by linear grammars, also contain some ambiguous languages.

Both finite and pushdown automata can be conveniently formalized using the notion of a rewriting system. We do not need such a formalization in what follows.

The *mirror image* $\bar{w}$ of a word w is obtained by writing w backwards. Each word vu is referred to a *circular variant* of a word $w = uv$. These notions are extended in the natural way to concern languages.

For a word w over the alphabet $\{a_1, \ldots, a_n\}$, we define its *Parikh vector* by

$$\psi(w) = (\#_{a_1}(w), \ldots, \#_{a_n}(w)).$$

(Recall that $\#_{a_i}(w)$ is the number of occurrences of a_i in w.) The *Parikh mapping* ψ is extended to concern languages:

$$\psi(L) = \{\psi(w) | w \in L\}.$$

The languages L_1 and L_2 are *Parikh equivalent* iff $\psi(L_1) = \psi(L_2)$. Sets $\psi(L)$ are referred to as *Parikh languages*.

Interconnections between languages and the corresponding Parikh languages are often useful in decidability considerations. We say that a set V of n-dimensional vectors with nonnegative integer entries is *linear* iff there are vectors $v_0, v_1, \ldots, v_k$ such that V consists of all vectors of the form

$$v_0 + \sum_{i=1}^{k} x_i v_i,$$

where x_i are nonnegative integers. A set V is *semi-linear* iff it is a finite union of linear sets. The term *"effectively"* has in this context the natural meaning: the "base" vector v_0 and the "periods" v_i, $i = 1, \ldots, k$, can be effectively computed from the given items, such as from the given context-free grammar.

Theorem 1.12. The Parikh language of a context-free language is effectively semilinear. The intersection of two given semilinear sets is effectively semilinear.

Only some part of the word under scan is rewritten at one step of the rewriting process defined above. According to *Lindenmayer systems* or *L systems*, rewriting happens in *parallel* in all parts of the word under scan. This reflects the original motivation: L systems were introduced to model the development of simple filamentous organisms. Context-free L systems are called *OL systems*. An OL system has at least one production for each letter of the alphabet. Every letter has to be rewritten during each step of the rewriting process. However, if the production $b \to b$ is in the set, then an occurrence of b can be kept unchanged. The deterministic version of an OL system, a *DOL system*, will be defined explicitly in Section 4.1. Several production sets, *"tables"*, are given in TOL and $DTOL$ *systems*. At each step of the rewriting process, only productions from the same table may be used. Of course, in the rewriting process according to a rewriting system, it does not make any difference whether we have one or more tables.

Post canonical systems will be needed in Chapter 2. The systems were originally introduced to model proofs in a formal mathematical system. The *rules* or *productions* in a Post canonical system are of the form

(*) $$(\alpha_1, \ldots, \alpha_k) \to \alpha ,$$

interpreted as "the conclusion α can be obtained from the premises $\alpha_1, \ldots, \alpha_k$". More explicitly, this means the following. Each of the αs is a word over $\Sigma_O \cup \Sigma_V$, where Σ_O and Σ_V are disjoint alphabets, consisting of *operating symbols* and *variables*. An *operating version*

$$(\alpha'_1, \ldots, \alpha'_k) \to \alpha'$$

of the rule (*) is obtained by substituting uniformly each letter of Σ_V with a word from Σ_O^*. (Here *uniform* substitution means that all occurrences of the same variable are replaced by the same word.) If we have already deduced each α'_i, $i = 1, \ldots, k$, we may deduce α'. For instance, the rule of *modus ponens* in logic can be written in these terms

$$(x , \ x \to y) \to y ,$$

where x and y are variables and $\to$ also denotes the logical implication.

The definition of a Post canonical system specifies a finite set of rules (*), as well as a finite set of *axioms*, words over Σ_O. Initially, we can use axioms as premises. Whenever we have inferred a word α', we can use it as a premise in later inferences, We still specify an alphabet $\Sigma_T \subseteq \Sigma_O$, the alphabet of *terminals*. The *language* *generated* by the system (also called the set of *theorems* in the system) consists of all words over Σ_T inferred in this fashion.

A *Post normal system* is a special type of Post canonical system: there is only one axiom and all rules are of the special form (with only one premise):

$$(**) \qquad \alpha X \to X\beta \, , \; \Sigma_V = \{X\} \, ; \; \alpha, \beta \in \Sigma_O^* \, .$$

It is obvious that Turing machines (viewed as rewriting systems) can be simulated by Post canonical systems. Consequently, the family of languages generated by Post canonical systems equals the family of recursively enumerable languages. It is less obvious that all canonical systems can be simulated by normal ones. The essential idea is to use the normal rules (**) to generate circular variants of words in such a way that various parts to be rewritten come to the beginning by turns. The classical reference [Min] contains a very good exposition of this construction. A variant of this construction will be met in Section 2.1. The reader is also referred to [Ho Ul], [Sal 1], [Sal 2] and [Ro Sa] for details concerning topics discussed in this section.

Theorem 1.13. A language is recursively enumerable iff it is generated by a Post normal system.

1.4 Creative sets and Gödel theory

We first return to sets of nonnegative integers and sharpen some of our earlier results. We then focus attention on *Gödel's Incompleteness Theory*. It will be seen that the issues involved bear an intimate relation to our undecidability considerations.

We will begin with some further facts concerning partial recursive functions. Our first aim is to formalize and discuss the informal claim that all general statements about partial recursive functions are undecidable.

We want to emphasize that an *equality between two partial functions* means that the functions are defined for the same arguments, and that the values are the same whenever the functions are defined. It will be clear from the context what the intended variables of the functions are. (Therefore, we do not introduce any special notation, such as the customary λ-notation, to indicate the variables.) As before, we consider a *fixed indexing* of partial recursive functions. If not stated otherwise, f_i is the partial recursive function of *one variable* computed by the Turing machine TM_i.

It is clear that our choice of indexing affects many properties of individual functions. For instance, f_{239} might be total in our indexing, but defined only for prime values of the argument in some other indexing. However, it is quite remarkable that some facts are independent of indexing. The most famous among them is the following fixed-point theorem, customarily called the *recursion theorem*.

Theorem 1.14. For every recursive function g, there is a natural number n (called the *fixed point* of g) such that

$$f_n = f_{g(n)}.$$

Proof. For each integer $i \geq 0$, we define a partial function H_i (of one variable) as follows. Given an argument x, TM_i is first applied to input i. (Thus, at this stage, the actual argument x is ignored!) If TM_i halts with i, producing output j, then we apply TM_j to input x. The output, if any, is the value $H_i(x)$. To summarize:

$$H_i(x) = \begin{cases} f_{f_i(i)}(x) & \text{if } f_i(i) \text{ converges}, \\ \text{undefined} & \text{if } f_i(i) \text{ diverges}. \end{cases}$$

Observe that $H_i(x)$ may be undefined even though $f_i(i)$ converges.

By Church's Thesis, every function H_i is partial recursive and, hence, possesses an index depending on i, say $h(i)$. But we have also given an algorithm for computing $h(i)$ given i. Consequently, another application of Church's Thesis tells us that h is recursive. Hence, we obtain

$$f_{h(i)}(x) = \begin{cases} f_{f_i(i)}(x) & \text{if } f_i(i) \text{ converges}, \\ \text{undefined} & \text{if } f_i(i) \text{ diverges}. \end{cases}$$

Consider now the given recursive function g. Clearly, the composition $g(h(x))$ is recursive and thus possesses an index k: $gh = f_k$. Moreover, gh is total, which implies that $f_k(k)$ converges. Altogether we obtain

$$f_{h(k)}(x) = f_{f_k(k)}(x) = f_{g(h(k))}(x).$$

This means that $n = h(k)$ is a fixed point, as required.

□

At this point, we could produce an unending list of undecidability results concerning partial recursive functions and/or recursively enumerable languages and sets. For instance, each of the following problems is undecidable:

(i) Is a given partial recursive function total? (In fact, if this problem were decidable, we could apply diagonalization to total functions alone.)
(ii) Is a given partial recursive function constant?
(iii) Is a given recursively enumerable language empty?
(iv) Is a given recursively enumerable language infinite?
(v) Is a given recursively enumerable language regular?
(vi) Do two given Turing machines accept the same language?
(vii) Do the ranges of two given partial recursive functions coincide?

Creative sets and Gödel theory

The proof of the undecidability of problems (i)–(vii) would be a rather straightforward matter at this stage and would just continue the argument begun in Theorems 1.2 and 1.3. We will not go into the details but will, rather, establish a general result: every nontrivial property of partial recursive functions or recursively enumerable languages is undecidable. Also, in this book we do not enter into discussion of *degrees* of undecidability: some problems are "more undecidable" than others. For example, the set of "no" instances of problem (iii) is recursively enumerable, whereas neither the set of "yes" instances nor the set of "no" instances of problem (iv) is recursively enumerable. (The set of "yes" instances can be identified as the set of indices i of all Turing machines TM_i accepting an infinite language.) In this sense, problem (iv) is "more undecidable" than (iii). In a certain well-defined sense, problem (v) is still more undecidable!

A property of partial recursive functions is called *nontrivial* iff it is possessed by at least one partial recursive function but not by all of them. Clearly, problems (i)–(vii) correspond to nontrivial properties. We now prove the general undecidability result, customarily referred to as *Rice's Theorem*. We formulate the result in terms of characteristic functions.

Theorem 1.15. Let F be a nonempty proper subset of partial recursive functions of one variable. Then the characteristic function of the set

$$S_F = \{i | f_i \in F\}$$

is not recursive.

Proof. Assume, on the contrary, that the characteristic function h of S_F is recursive. We compute the values $h(0)$, $h(1)$, ... until we find numbers j and k such that $h(j) = 1$ and $h(k) = 0$, which means that

$$f_j \in F \quad \text{and} \quad f_k \notin F.$$

By our assumptions concerning F, numbers j and k will eventually be found.

Let h' be any recursive function mapping 0 to j and 1 to k. Consequently, the composition $g = h'h$ is a recursive function satisfying

$$g(x) = \begin{cases} k & \text{if } f_x \in F, \\ j & \text{if } f_x \notin F. \end{cases}$$

We apply Theorem 1.14 to g, and consider a fixed point n with the property

$$f_n = f_{g(n)}.$$

We are now in trouble with index n. Assume, firstly, that f_n is in F. Hence, $g(n) = k$. But $f_k = f_n$ is not in F! Assume, secondly, that f_n is not in F. Hence $g(n) = j$. But, again by $f_n = f_{g(n)}$, we infer that $f_n = f_j$ is in F! Thus, a contradiction always arises.

□

Recall that every partial recursive function has infinitely many indices. If we consider a partial recursive function, we are at the same time considering an infinite index set for the function. Undecidability proofs make use of this state of affairs. On the other hand, nontrivial properties of indices, that is, nontrivial properties of Turing machines might very well be decidable. For instance, each of the following nontrivial properties of Turing machines is easily shown to be decidable. Has a given Turing machine exactly three states? Does a given Turing machine move to the right all the time when started with the input 0? Does a given Turing machine visit at least one square at least 100 times when started with the input 0? It is also easy to find indices i such that the halting problem of TM_i is decidable: given an input x, we can decide whether or not TM halts with x.

We now introduce two further notions, those of *productive* and *creative sets*. There are many interconnections with decidability and logic in general. The term "creative" was originally introduced by Post to reflect the fact that provable statements in a sufficiently strong formal system form a creative set.

In Theorem 1.3 we considered the characteristic function of the set

$$SA = \{i | f_i(i) \text{ converges }\}.$$

We know that SA is not recursive. Since SA is recursively enumerable, $\sim SA$ cannot be recursively enumerable. In fact, we can make a stronger claim. Whenever S_i is a recursively enumerable set (that is, S_i is the domain of the partial recursive function f_i) contained in $\sim SA$, then the index i belongs to the difference $\sim SA - S_i$. In particular, $\sim SA$ itself is not recursively enumerable because then the difference would be empty.

The proof of the claim $i \in \sim SA - S_i$ is immediate. The inclusion $S_i \subseteq \sim SA$ implies that, whenever x is in S_i, then $f_x(x)$ diverges. Consequently, $f_i(i)$ diverges, which, together with the inclusion, proves the claim.

We can also express the claim in the following more general form. Whenever a recursively enumerable set S_i is contained in $\sim SA$, then we can effectively find a number in the difference $\sim SA - S_i$. This is the idea behind the following definition of productive sets.

A set S of nonnegative integers is termed *productive* iff there is a partial recursive function f such that, for all i, the inclusion $S_i \subseteq S$ implies that $f(i)$ is defined and belongs to the difference $S - S_i$. A recursively enumerable set is termed *creative* iff its complement is productive.

Thus, $\sim SA$ is productive and SA is creative. No creative set is recursive. Thus, we have exhibited two disjoint subclasses of the class of recursively enumerable sets: recursive sets and creative sets. The two subclasses by no means cover the whole class of recursively enumerable sets, but the details of this matter lie beyond the scope of this book.

We will now present some basics of Gödel's celebrated incompleteness theory, particularly from the point of view of the theory of recursive and recursively enumerable sets. We consider some *formal mathematical system*. (We also speak

Creative sets and Gödel theory

interchangeably of an *axiomatic system* or of a *formal theory*.) The system could be some formalization of arithmetic or, for instance, some axiomatic system of set theory including formalized arithmetic.

Example 1.7. In the formalization of arithmetic we have to specify first a language that is adequate for making statements concerning arithmetic of nonnegative integers. The alphabet includes the symbols 0, 1, +, × (times), =, variables, parentheses, propositional connectives and quantifiers. *Well-formed formulas* of the language are defined inductively in a rather obvious fashion — we omit the details. For instance,
$$\exists y((1+1) \times y = z)$$
is the formal counterpart of the statement "z is even". When the statements become more complicated, various notational conventions and abbreviations will be necessary. For instance, we abbreviate $1+1$ by 2, $2+1$ by 3, and so forth. For the formula
$$\exists z(\neg(z=0) \wedge (x = y + z))$$
we use the abbreviation $x > y$. Instead of $\neg(x=0)$ we apply the customary $x \neq 0$.

This should give some idea of how various statements about nonnegative integers can be represented in formalized arithmetic. Let us write the statement "there are infinitely many twin primes". (Twin primes are pairs of primes whose difference is 2, such as 41 and 43.) We abbreviate the formula
$$(x \neq 0) \wedge (x \neq 1) \wedge \forall y \forall z\, (x = y \times z \to ((y=1) \vee (z=1)))$$
as $PRIME(x)$. Then the statement mentioned can be written
$$(\forall x)\,((PRIME(x) \wedge PRIME(x+2))$$
$$\to (\exists y)((y > x) \wedge PRIME(y) \wedge PRIME(y+2)))\,.$$

We come next to the concept of a *proof* in formalized arithmetic and to the set of *theorems*, that is, provable formulas. A proof should establish the validity of its conclusion, given the validity of its assumptions. To establish the validity of the conclusion convincingly, the proof must be verifiable. There must be an effective procedure for verifying the correctness of every step in the proof. In a formal theory, such as formalized arithmetic, some statements are specified as *axioms*. The set of axioms must be recursive. Each axiom is a theorem, and new theorems are obtained by applying specified *rules of inference* to previously established theorems. Usually there are only finitely many rules of inference. The verification of a step means that one studies the statement under scan and checks that it is either an axiom or follows from earlier steps in the proof by some rule of inference. Thus, there is an algorithm for verifying whether or not a given finite sequence of formulas (statements) is a proof. In other words, the set of proofs is recursive. We can also express this by saying that the predicate $PROOF(x, y)$ — x is a finite sequence of statements, the last one of which is y, and x is a proof of y

— is decidable or recursive. Finally, the predicate $THEOREM(y)$, y is a theorem, can be defined as $(\exists x) PROOF(x, y)$, and hence it is recursively enumerable. If $THEOREM(y)$ is valid, we also say that y is *provable* and use the notation $\vdash y$. The symbol $\vdash$ may also be provided with an index to indicate the formal system we are working with. Similarly, y is *refutable* iff its negation is provable, that is, $\vdash \neg y$. The formal system is *complete* iff every (well-formed) formula is provable or refutable. It is *consistent* iff there is no formula y such that both $\vdash y$ and $\vdash \neg y$.

We hope that this brief and sketchy description will serve its purpose as an introduction to the basics of Gödel's theory discussed below.

□

Using a procedure similar to that used to enumerate Turing machines and partial recursive functions, we can enumerate statements or well-formed formulas in a formal system. The procedure associates to each formula y a nonnegative integer $gn(y)$, referred to as the *Gödel number* or *index* of y. Let us assume that such an indexing has been performed. The details are largely irrelevant. However, the transition from y to $gn(y)$, and vice versa, must be effective.

We can now associate concepts, defined originally for sets of nonnegative integers, to sets of formulas in the natural fashion. We say that a set F of formulas is *recursive* iff the set

$$gn(F) = \{gn(y) | y \in F\}$$

is recursive. (For instance, F might be the set of provable or refutable formulas.) Similarly, we define what it means for F to be *recursively enumerable, productive* or *creative*.

From now on we consider a fixed formal mathematical system AX that is strong enough to include the arithmetic (addition and multiplication) of nonnegative integers. We assume that AX is *consistent*. For instance, AX can be the well-known *Peano arithmetic* or some consistent extension of it. The details of the system AX are irrelevant for us because we do not enter into the constructions needed for the proof of Lemma 1.17 below. Provability within AX is denoted by the symbol $\vdash$.

We now introduce some modifications of the "self-applicability set"

$$SA = \{i | f_i(i) \text{ converges}\}$$

discussed earlier. Consider the following two subsets of SA:

$$SA_0 = \{i | f_i(i) = 0\} \quad \text{and} \quad SA_1 = \{i | f_i(i) = 1\}.$$

Clearly, SA_0 and SA_1 do not intersect. It is also obvious that SA_0 and SA_1 are recursively enumerable.

Two sets S and S' with an empty intersection are called *recursively inseparable* iff there is no recursive set R such that

$$S \subseteq R \quad \text{and} \quad S' \subseteq \sim R.$$

Creative sets and Gödel theory 43

It follows that if S and S' are recursively inseparable then neither one of them is recursive.

Theorem 1.16. SA_0 and SA_1 are recursively inseparable.

Proof. We have already observed that the relation $SA_0 \cap SA_1 = \emptyset$ follows by the definition of SA_0 and SA_1. We proceed indirectly and assume that a recursive set R separates SA_0 and SA_1:

$$(*) \qquad SA_0 \subseteq R \quad \text{and} \quad SA_1 \subseteq \sim R.$$

Let $h = f_j$ be the characteristic function for R. Hence, h is a total function, always assuming the value 0 or the value 1.

What about the value $h(j) = f_j(j)$? If $h(j) = 0$, then $j \in \sim R$ and hence, by $(*)$, $h(j) = 1$. If $h(j) = 1$, then $j \in R$. Also, $(*)$ now gives the contradictory result $h(j) = 0$. Consequently, there is no recursive R satisfying $(*)$. □

We now associate to the set SA_0 the following binary predicate $P_0(i, c)$. For a pair (i, c), $P_0(i, c)$ holds iff $f_i(i) = 0$ and c is the index (Gödel number) of the computation of the Turing machine TM_i, showing that TM_i produces output 0 when started with input i. Halting computations are finite sequences of instantaneous descriptions and, as such, can be represented by number c. (As always, we assume that the transition from a computation to the number c, as well as from c to the computation, is effective.) Clearly, $P_0(i, c)$ is a decidable predicate. Moreover, the membership $i \in SA_0$ can be expressed as $(\exists c) P_0(i, c)$.

There is a well-formed formula $(\exists c)\alpha_0(i, c)$ in the language of AX such that, whenever $i \in SA_0$, then $\vdash (\exists c)\alpha_0(i, c)$. We cannot prove this without entering into the details of AX — and such details are not important for our main purposes. The basic idea is that the fact of a computation step being correct can be expressed as an arithmetical fact between the encodings of two instantaneous descriptions. The latter fact, in turn, can be established as a formalized proof in AX. If performed in detail, this is a very lengthy and tedious task: the numeral i appearing in the formula must first be used to generate TM_i, after which legal computation steps can be formalized within AX. Some details of a similar construction will be given at the end of Chapter 2.

The set SA_1 is handled similarly. We conclude that there is a well-formed formula $(\exists c)\alpha_1(i, c)$ such that, whenever $i \in SA_1$, then also $\vdash (\exists c)\alpha_1(i, c)$.

We now introduce a further trick, known as *Rosser's modification*. Denote by $\beta_0(i)$ (resp. $\beta_1(i)$) the formula

$$(\exists c)\alpha_0(i, c) \land \forall x(x \leq c \to \neg \alpha_1(i, x))$$

$$(\text{resp. } (\exists c)\alpha_1(i, c) \land \forall x(x \leq c \to \neg \alpha_0(i, x))) \, .$$

It is straightforward to establish, by the consistency of AX, that $\vdash \beta_0(i)$ follows from $\vdash (\exists c)\alpha_0(i, c)$. (Indeed, the second term of the conjunction is a formal consequence from the first term by consistency. This would not be true for unbounded universal quantification; then the second term would contain conceivably unprovable information.) A similar conclusion can be made with regard to $\beta_1(i)$. Moreover, from $\vdash \beta_1(i)$ we also obtain $\vdash \neg\beta_0(i)$.

To summarize, we have the following result.

Lemma 1.17. (i) If $i \in SA_0$, then $\vdash \beta_0(i)$. (ii) If $i \in SA_1$, then $\vdash \beta_1(i)$. (iii) If $\vdash \beta_1(i)$, then $\vdash \neg\beta_0(i)$.

We are now ready to establish the main result, customarily referred to as *Rosser's modification of Gödel's First Incompleteness Theorem* or the *Gödel–Rosser Incompleteness Theorem*.

Theorem 1.18. Assuming that AX is consistent, there is a formula α (referred to as a *Gödel sentence* for AX) such that neither $\vdash \alpha$ nor $\vdash \neg\alpha$.

Proof. We consider the sets of nonnegative integers

$$PR = \{i | \vdash \beta_0(i)\} \quad \text{and} \quad REF = \{i | \vdash \neg\beta_0(i)\}.$$

The consistency of AX implies that

$$PR \cap REF = \emptyset.$$

Lemma 1.17 (i) gives the inclusion $SA_0 \subseteq PR$. Lemma 1.17 (ii)–(iii) gives the inclusion $SA_1 \subseteq REF$. We now use an idea based on the recursive inseparability (see Theorem 1.16) of the sets SA_0 and SA_1.

Consider the partial function

$$g(i) = \begin{cases} 1 & \text{if } i \in PR, \\ 0 & \text{if } i \in REF, \\ \text{undefined} & \text{otherwise}. \end{cases}$$

Because the sets PR and REF are obviously recursively enumerable (since we can enumerate the proofs of AX), we conclude that g is a partial recursive function and, hence, possesses an index $j : g = f_j$.

We now claim that $\beta_0(j)$ qualifies as a Gödel sentence. Assume first that $\vdash \beta_0(j)$. This implies that $j \in PR$ and, hence, $g(j) = f_j(j) = 1$. But, by the definition of SA_1, this implies that $j \in SA_1$. Since $SA_1 \subseteq REF$, we obtain the result $j \in REF$, which means that $\vdash \neg\beta_0(j)$. Because AX is consistent, we cannot have both $\vdash \beta_0(j)$ and $\vdash \neg\beta_0(j)$. This means that our original assumption $\vdash \beta_0(j)$ is wrong.

Assume, secondly, that $\vdash \neg\beta_0(j)$. The reasoning in this case proceeds through the following conclusions:

$$\begin{aligned} j &\in REF, \\ g(j) &= f_j(j) = 0, \\ j &\in SA_0, \\ j &\in PR. \end{aligned}$$

Creative sets and Gödel theory

We need not go further because $REF \cap PR = \emptyset$. We conclude that the assumption $\vdash \neg \beta_0(j)$ is wrong. We have thus shown that $\beta_0(j)$ is a Gödel sentence for AX. □

In his original formulation Gödel assumed that AX is ω-*consistent*: there is no formula $\alpha(i)$ such that all of the following are provable:

$$(\exists i)\alpha(i) \, , \, \neg\alpha(0) \, , \, \neg\alpha(1) \, , \, \neg\alpha(2) \, , \ldots \, .$$

This assumption is stronger than the assumption of consistency. Assuming ω-consistency, we can work with the formula $(\exists c)\alpha_0(i,c)$ rather than $\beta_0(i)$ in the construction of a Gödel sentence. The advantage gained in Rosser's modification is that we obtain point (iii) in Lemma 1.17.

The only requirement for AX is that Lemma 1.17 can be established. Then the proof of Theorem 1.18 can be carried out. There is no possibility of avoiding incompleteness by adding new axioms. Using the recursive inseparability of the sets PR and REF, it is fairly straightforward to show that the set of formulas provable in AX is creative.

We mention, finally, *Gödel's Second Incompleteness Theorem*. The consistency of AX can be expressed by saying that the formula $0 = 1$ is not provable. (If it were provable, then all well-formed formulas would be provable by rules of logic.) On the other hand, the negation $\neg(\exists x)PROOF(x, 0 = 1)$ can be represented in AX by a formula; denote it by CON. The proof of the next theorem, Gödel's Second Incompleteness Theorem, can be carried out by showing how a contradiction arises from a proof of CON. The reader is referred, for instance, to [Bör] for further details.

Theorem 1.19. *If AX is consistent, then not $\vdash CON$.*

Interlude 1

Bolgani. Before we go into any other problems, let me solve a problem concerning *Dramatis personae*. So this remark of mine belongs to meta-drama. The voices of David, Kurt and Emil come from [Hil], [Dav] and [Göd 2]. The order of some sentences has been changed, and some idioms have been modernized in translations from German. Thus, the term "complexity measure" is used although "criterion for simplicity" would have been a linguistically more correct translation of David's German words. This happens very infrequently, though.

Tarzan. I observed something like this already before when you made a reference to [Göd 1], one of the most famous scientific papers in this century. I have now read through Chapter 1. I fully realize the necessity of staying at an informal or semiformal level. Fully formal constructions would not only be tedious but also unreadable.

Emil. In the old mathematics the formal and informal were confused, in symbolic logic the informal tended to be neglected. Yet the real mathematics must lie in the informal development. For in every instance the informal "proof" was first obtained; and once obtained, transforming it into the formal proof turned out to be a routine chore. Our present formal proofs, while complete, will require drastic systematization and condensation prior to publication.

Tarzan. I also compared the finite 1-process discussed in the Prologue with the Turing machine. To "set up" a finite 1-process amounts to defining a Turing machine, the tape and the instructions of the Turing machine being the "symbol space" and the "directions" of the 1-process.

Bolgani. And both concepts, the Turing machine and Post's 1-process, were thought of simultaneously and independently of each other, as well as independently of any previous mathematical model or formalization of effective procedures. This is true of many of the alternate formalizations listed in (i) in the evidence supporting Church's Thesis, presented in Section 1.1. One gets the impression that we are dealing here with a *law of nature* that has been observed from various angles but still remains the same. For instance, we can define the notion of a "Post 1-function" and demonstrate, with some effort, that this notion leads to the class of computable functions as defined in Section 1.1.

Emil. The purpose of the formulation of 1-processes is not only to present a system of a certain logical potency but also, in its restricted field, of psychological fidelity. In the latter sense wider and wider formulations are contemplated. Our aim will be to show that all such are logically reducible to the formulation of 1-processes. We offer this conclusion at the present moment as a *working hypothesis*. And to our mind such is Church's identification of effective calculability with recursiveness.

Bolgani. This refers to Church's Thesis.

Emil. Out of this working hypothesis a Gödel-Church development flows independently. The success of our program would, for us, change this hypothesis not so much to a definition or to an axiom as to a *natural law*. Only thus can Gödel's theorem concerning the incompleteness of symbolic logics of a certain general type and Church's results on the recursive unsolvability of certain problems be transformed into conclusions concerning all symbolic logics and all methods of solvability.

Tarzan. I understand that by now the program has had all the success one could hope for. We can think of Church's Thesis as a natural law. It was used as a very powerful methodological tool in many proofs of Chapter 1; just remember how easily we obtained the universal Turing machine. There was no need for involved subroutines or complicated constructions. And, as predicted by Emil, the conclusions concern all formal systems and all ideas of undecidability.

Kurt. The great importance of the concept of computability is largely due to the fact that with this concept one has for the first time succeeded in giving an absolute definition of an interesting epistemological notion, i.e., one not depending on the formalism chosen. In all other cases treated previously, such as demonstrability or definability, one has only been able to define them relative to a given language, and for each individual language it is clear that the one thus obtained is not the one looked for. For the concept of computability, however, although it is merely a special kind of demonstrability or decidability, the situation is different. By a kind of miracle it is not necessary to distinguish orders, and the diagonal procedure does not lead outside the defined notion. Thus the notion of "computable" is in a certain sense "absolute", while almost all meta-mathematical notions otherwise known (for example, provable, definable, and so on) depend essentially upon the system adopted.

Bolgani. There is also something absolute in the central idea of Gödel-numbering or indexing: it does not matter how we do it, provided the index of an item (whatever it may be) is computable and, conversely, the item is computable from the index.

Emil. The "Gödel Representation" is merely a case of *resymbolization*. With the growth of mathematics such resymbolization will have to be effected again and again.

Tarzan. The notion of a formal mathematical system or an axiomatic system seems to be pretty well understood nowadays, and the details of the definition seem to be uniformly accepted. It would be interesting to get a rough idea as to how things were initially formulated in the 1930s.

Kurt. A *formal mathematical system* is a system of symbols together with the rules for employing them. The individual symbols are called *undefined terms*. *Formulas* are certain finite sequences of undefined terms. A class of formulas called *meaningful formulas*, and a class of meaningful formulas called *axioms* will be defined. There may be a finite or an infinite number of axioms. Further, a list of rules will be specified, called *rules of inference*; if such a rule is called R, it defines the relation of *immediate consequence by R* between a set of meaningful formulas $M_1, \ldots, M_k$, called the *premises*, and a meaningful formula N, called the *conclusion* (ordinarily $k = 1$ or 2). We require that the rules of inference, and the definitions of meaningful formulas and axioms, be constructive; that is, for each rule of inference there shall be a finite procedure for determining whether a given formula B is an immediate consequence (by that rule) of given formulas $A_1, \ldots, A_n$, and there shall be a finite procedure for determining whether a given formula A is a meaningful formula or an axiom.

A formula N shall be called an *immediate consequence* of $M_1, \ldots, M_n$ if N is an immediate consequence of $M_1, \ldots, M_n$ by any one of the rules of inference. A finite sequence of formulas shall be a *proof* (specifically, a proof of the last formula of the sequence) if each formula of the sequence is either an axiom, or an immediate consequence of one or more of the preceding formulas. A formula is *provable* if a proof of it exists.

Bolgani. The decidability issues are taken care of here. It is decidable whether or not a given sequence of formulas is a proof. In the terminology of Chapter 1, the set of provable formulas is recursively enumerable.

David. Our thinking is finitary; while we think, a finitary process is happening. My proof theory makes use of this self-evident truth to a certain extent in the following way. If there were a contradiction somewhere, then for the recognition of this contradiction the corresponding choice from the infinitely many choices would also have to be made. Accordingly, my proof theory does not claim that the choice of an object from infinitely many objects could always be made. The claim is rather that one can always proceed, without any risk of error, as if the choice had been made. Here the application of *tertium non datur* is without danger.

In my proof theory the finite axioms are supplemented by transfinite axioms and formulas, similarly as imaginary elements are added to the real ones in the theory of complex numbers, or ideal structures to the real ones in geometry. And similarly as in these other areas, this method also leads to success in my proof theory: the

Interlude 1 49

transfinite axioms at the same time simplify and close the theory.

Tarzan. Here the decidability issues are not taken care of. One almost gets the feeling that experimental verification in some world, real or imaginary, is called for.

Bolgani. The difference between the two notions become much more clearly understood later. I mean the difference concerning whether or not the rules of the game are decidable. I think the matter can be summarized in an excellent way.

Kurt. As a consequence of later advances, in particular of the fact that, due to A. M. Turing's work, a precise and unquestionably adequate definition of a formal system can now be given, the existence of undecidable arithmetical propositions and the nondemonstrability of the consistency of a system in the same system can now be proved rigorously for *every* consistent formal system containing a certain amount of finitary number theory.

Turing's work gives an analysis of the concept of "mechanical procedure" (alias "algorithm" or "computation procedure" or "finite combinatorial procedure"). This concept is shown to be equivalent to that of a "Turing machine". (The paper by E. L. Post [Post 1] was almost simultaneous.) A *formal system* can simply be defined to be any mechanical procedure for producing formulas, called provable formulas. For any formal system in this sense there exists one in the sense I described earlier that has the same provable formulas (and likewise vice versa), provided the term "finite procedure" that I used earlier is understood to mean "mechanical procedure". This meaning, however, is required by the concept of a formal system, whose essence it is that reasoning is completely replaced by mechanical operations on formulas. (Note that the question of whether there exist finite *nonmechanical* procedures, not equivalent with any algorithm, has nothing whatsoever to do with the adequacy of the definition of "formal system" and of "mechanical procedure".)

Chapter 2

Post Correspondence Problem

2.1 What Emil said

The choice of a suitable reference point is important in establishing the undecidability of a given problem P. The reference point R is known to be undecidable. The problem P is shown to be undecidable by the indirect reduction argument: if P were decidable then R would also be. The *Post Correspondence Problem* is the most suitable reference point for undecidability proofs in language theory, the reason being that it captures the essence of grammatical derivations.

The Post Correspondence Problem, first defined by Post in [Post 2], can be described simply as follows. We are given two lists, both consisting of $n \geq 1$ words over an alphabet Δ. We have to decide whether or not some catenation of words in one list equals the catenation of the *corresponding* words in the other list. This means that words in exactly the same positions in the two lists must be used. If the first, third and twice the first word were chosen from one list in this order, these four words would also have to be chosen from the other list in the same order.

Example 2.1. Let the two lists be

$$(a^2, b^2, ab^2) \text{ and } (a^2b, ba, b).$$

Catenate the first, second, first and third word. The word $a^2b^2a^3b^2$ results from both lists. Thus, in this case we find a catenation of the required type. We express this by saying that this *instance PCP* of the Post Correspondence Problem has a *solution* 1213. Clearly, solutions can be catenated to yield another solution. Consequently, all words of the form $(1213)^i$, $i \geq 1$, over the "index alphabet" $\Sigma = \{1, 2, 3\}$ are solutions.

□

Two facts can be learned from this example. If we are on the way to a solution, one of the two words under scan must be a prefix of the other. If an instance PCP has a solution, it has infinitely many of them. A solution w is termed *primitive* if no proper prefix of w is a solution. In the preceding example 1213 is a primitive solution.

Example 2.2. The instance defined by the two lists (a, b^2) and (a^2, b) has no solution. If a candidate for a solution begins with index 1 (resp. 2), the next indices also have to be 1s (resp. 2s). No solution results because one of the resulting words will always be a proper prefix of the other.

□

Example 2.3. Also, the instance defined by the two lists (a^2b, a) and (a^2, ba^2) possesses no solution. This can be seen as follows. An eventual solution must begin with index 1. Index 2 must follow because the word coming from the second list must "catch up" the missing b. So far, we have the words a^2ba and a^2ba^2. Since one of the words must always be a prefix of the other, we conclude that the next index in the eventual solution must be 2, yielding the words a^2ba^2 and $a^2ba^2ba^2$. No further continuation is possible because both words in the first list begin with a.

□

Example 2.4. A similar argument shows that the instance with lists (a^2b, a^2) and (a, ba^2) has no solution. On the other hand, the slightly modified instance with lists (a^2b, a) and (a, ba^2) has the solution 1122. A simple analysis shows that it is the only primitive solution. It is also easy to see that 1213 is the only primitive solution in Example 2.1.

□

The Post Correspondence Problem can be defined in a compact way using morphisms.

Definition. Let g and h be two morphisms mapping Σ^* into Δ^*. The pair $PCP = (g, h)$ is called an *instance* of the *Post Correspondence Problem*. A nonempty word w over Σ such that $g(w) = h(w)$ is referred to as a *solution* of PCP. A solution w is *primitive* if no proper prefix of w is a solution.

Clearly, the two lists of words we were talking about before are obtained by listing the values $g(a)$ and $h(a)$, where a ranges over Σ. In this sense Σ can be viewed as the alphabet of indices.

A further remark concerning primitive solutions is in order. Primitivity means *suffix primitivity*: no proper suffix can be removed from a solution and still obtain a solution. One can also require that no proper subword can be removed and still obtain a solution. Solutions satisfying this requirement are referred to as *subword primitive*. *Scattered primitive* solutions are defined by the condition that no scattered subword can be removed and still obtain a solution. Clearly, every scattered primitive solution is subword primitive, and every subword primitive solution is suffix primitive.

Example 2.5. The PCP defined by the morphisms

$$g: \quad 1 \to b^2 \;, \quad 2 \to ab \;, \quad 3 \to b$$
$$h: \quad 1 \to b \;, \quad 2 \to ba \;, \quad 3 \to b^2$$

is from Post's original [Post 1]. The solutions 12^i3, $i \geq 1$, are suffix primitive but not subword primitive, since 13 is also a solution. The PCP defined by the morphisms

$$g: \quad 1 \to b^2 \;, \quad 2 \to b^3 \;, \quad 3 \to b \;, \quad 4 \to b$$
$$h: \quad 1 \to b \;, \quad 2 \to b \;, \quad 3 \to b^2 \;, \quad 4 \to b^3$$

has the solution 1234 that is subword primitive but not scattered primitive, since 13 is also a solution. □

The examples given above show that, for many instances PCP, solvability is easy to settle. We will see below in Section 2.3 that the same also holds for some infinite classes of instances. However, it is shown in the next theorem that no algorithm for solving instances of special types can be extended to take care of *every* instance.

Customary proofs of Theorem 2.1 are based on a reduction to the halting problem or to the membership problem of recursively enumerable languages. (For instance, see [Sal 1] or [Ho Ul].) This leads to some rather awkward technical difficulties if one wants to do a good job in the details. We will follow the original approach of Post [Post 2]. It will become apparent why the Correspondence Problem arises very naturally in the setup of normal systems. Perhaps this also explains the title "A variant of a recursively unsolvable problem" of [Post 2]. It is difficult to imagine a more modest title for a paper as influential as this one has been!

Theorem 2.1. The Post Correspondence Problem is undecidable.

Proof. Consider a normal system S with axiom u and rules

$$\alpha_i X \to X \beta_i \;, \quad i = 1, \ldots, n \;,$$

where u, α_i and β_i are words over the alphabet Δ. By Theorem 1.13, the membership problem of $L(S)$ is undecidable. Starting from the pair (S, v), where v is an arbitrary word over Δ, we will construct an instance PCP of the Post Correspondence Problem such that PCP possesses a solution iff v is in $L(S)$. Thus, the decidability of the Post Correspondence Problem would contradict Theorem 1.13. The instance PCP is closely related to the instance determined by the two lists α_i and β_i, $i = 1, \ldots, n$.

As an illustration, consider the example due to Post, already discussed in Example 2.5 above. Let $\Sigma = \{1, 2, 3\}$, $\Delta = \{a, b\}$, and

$$g(1) = b^2 = \alpha_1 \;, \quad g(2) = ab = \alpha_2 \;, \quad g(3) = b = \alpha_3 \;,$$
$$h(1) = b = \beta_1 \;, \quad h(2) = ba = \beta_2 \;, \quad h(3) = b^2 = \beta_3 \;.$$

Choosing $n = 3$, the αs and βs as above and $u = abab^3$, we obtain a normal system S. The derivation

$$abab^3 \Rightarrow ab^4a \Rightarrow b^3aba \Rightarrow b^2abab^2$$

shows that the word $v = b^2abab^2$ is in $L(S)$. The following equations can be seen from this derivation:

$$u = \alpha_2 ab^3 \, , \ ab^3\beta_2 = \alpha_2 b^3 a \, , \ b^3 a \beta_2 = \alpha_3 b^2 aba \, , \ b^2 aba\beta_3 = v \, .$$

The procedure is always the following. The prefix α_i is removed and the suffix β_i is added. If the derivation continues, the resulting word again has some α_j as a prefix. This shows that

$$u\beta_2\beta_2\beta_3 = \alpha_2\alpha_2\alpha_3 v \, .$$

Exactly the same argument applies for an arbitrary normal system S, defined as above. Assume that a word v has a derivation, where the rules with indices $i_1, i_2, \ldots, i_k$ are applied in this order. The derivation yields the equations

$$u = \alpha_{i_1} x_1 \, , \ x_1 \beta_{i_1} = \alpha_{i_2} x_2 \, , \ldots, \ x_{k-1}\beta_{i_{k-1}} = \alpha_{i_k} x_k \, , \ x_k \beta_{i_k} = v \, ,$$

for some words (possibly empty) x. The equations, in turn, yield the single equation

(*) $\qquad u\beta_{i_1}\beta_{i_2} \ldots \beta_{i_k} = \alpha_{i_1}\alpha_{i_2} \ldots \alpha_{i_k} v \, .$

If we denote the word $i_1 i_2 \ldots i_k$ by w and consider the morphisms g and h defined as above, the equation can be written as

$$uh(w) = g(w)v \, .$$

To summarize: if v is in $L(S)$ then this equation holds for some word w over the alphabet $\Sigma = \{1, 2, \ldots, n\}$.

The converse does not hold true. This is due to the fact that we obtain, in a manner similar to (*), the equations

$$u\beta_{i_1} \ldots \beta_{i_{j-1}} = \alpha_{i_1} \ldots \alpha_{i_j} x_j \, , \ j = 1, \ldots, k \, ,$$

and hence,

(**) $\qquad |u\beta_{i_1} \ldots \beta_{i_{j-1}}| \geq |\alpha_{i_1} \ldots \alpha_{i_j}| \, , \ j = 1, \ldots, k \, .$

Consequently, if v is in $L(S)$, then there is a word $w = i_1 i_2 \ldots i_k$ such that (*) and (**) hold. Conversely, given such a word w, the x_j can be determined and, hence, the derivation of v can be constructed. This implies that (*) *and* (**) *constitute a necessary and sufficient condition for v being in $L(S)$*.

It is easy to give examples of normal systems, where (*) does not imply (**). For instance, the choice

$$u = a \, , \ v = \lambda \, , \ \alpha_1 = a^2 \, , \ \beta_1 = a$$

gives $u\beta_1 = \alpha_1 v$ but because $|u| < |\alpha_1|$, the equation cannot be converted into a derivation. Clearly, v is not derivable from u. We may consider the additional pair $\alpha_2 = \beta_2 = a$ and obtain, for any i, $u\beta_2^i\beta_1 = \alpha_2^i\alpha_1 v$ but no derivation of v

results, although we now obtain an initial part of a derivation as long as we wish. However, we always meet a failure of (**), which means that there is no space for the further choice α_{i_j} and, hence, for a further derivation step.

Our *first task* is to show how our arbitrary S can always be replaced by another normal system, where (*) implies (**). Our *second task* is to eliminate u and v, after which we only have to deal with an equation of the form $g(w) = h(w)$, i.e., a solution of PCP.

Consider the first task. Recall that the *mirror image* of a word α is denoted by $\bar{\alpha}$. Replace the axiom u of S by $\bar{u}$ and the rules by

$$X\overline{\alpha_i} \to \overline{\beta_i}X \, , \quad i = 1, \ldots, n \, .$$

The language $L(S')$ of the resulting Post system S' (not normal) consists of the mirror images of the words in $L(S)$. This is immediately seen by induction, using the formula $\overline{\alpha\beta} = \bar{\beta}\bar{\alpha}$.

Add a new letter $\#$ to the alphabet Δ of S and S'. Let S'' be the Post system with the axiom $\bar{u}\#$ and rules

$$X\overline{\alpha_i}\# \to \overline{\beta_i}X\# \, , \quad i = 1, \ldots, n \, .$$

The derivations according to S'' are exactly the same as those according to S', except that the right end marker $\#$ has been added. Consequently,

$$L(S'') = \{\bar{v}\# \mid v \text{ in } L(S)\} \, .$$

Consider, finally, the *normal* system S''' with the axiom $\bar{u}\#$ and rules

$$\overline{\alpha_i}\#X \to X\#\overline{\beta_i} \, , \quad i = 1, \ldots, n \, ,$$
$$\#X \to X\# \, , \quad aX \to Xa \, , \quad a \in \Delta \, .$$

The rules in the second line serve the purpose of transferring an arbitrary prefix of a word over $\Delta \cup \{\#\}$ to be a suffix. With this in mind, we see that the rules in the first line simulate the rules of S'' and, conversely, the rules of S'' simulate the rules in the first line. More specifically,

$$\bar{v}\# \text{ is in } L(S''') \text{ iff } \bar{v}\# \text{ is in } L(S'') \text{ iff } v \text{ is in } L(S) \, .$$

Observe that $L(S'')$ is properly contained in $L(S''')$. In fact, $L(S''')$ consists of all of the "circular variants" of the words in $L(S'')$.

Rename the rules of S''' as

$$\gamma_i X \to X\delta_i \, , \quad i = 1, \ldots, n + \text{card}(\Delta) + 1 \, .$$

Observe that none of the words γ_i and δ_i is empty. Assume that, for some word $w = i_1 i_2 \ldots i_k$,

$$(*)_1 \qquad \bar{u}\#\delta_{i_1}\delta_{i_2}\ldots\delta_{i_k} = \gamma_{i_1}\gamma_{i_2}\ldots\gamma_{i_k}\bar{v}\# \, .$$

We claim that

$(**)_1$ $\qquad |\bar{u}\#\delta_{i_1}\ldots\delta_{i_{j-1}}| \geq |\gamma_{i_1}\ldots\gamma_{i_j}|$, $j = 1,\ldots,k$.

Let g_1 and h_1 be the morphisms defined by

$$g_1(i) = \gamma_i, \quad h_1(i) = \delta_i, \quad i = 1,\ldots,n + \text{card}(\Delta) + 1.$$

If our claim is true, then in view of our earlier discussion we have reached the following subgoal.

Subgoal. A word v is in $L(S)$ iff, for some word w,

$(*)_2$ $\qquad \bar{u}\#h_1(w) = g_1(w)\bar{v}\#$.

To establish our claim, assume the contrary: $(*)_1$ holds but $(**)_1$ fails for a specific j. This implies that

$$\gamma_{i_1}\ldots\gamma_{i_j} = \bar{u}\#\delta_{i_1}\ldots\delta_{i_{j-1}}y,$$

for some nonempty word y. The rules of S''' neither introduce nor eliminate occurrences of $\#$. The word γ_{i_j} is either a letter of Δ or else ends with $\#$ and has no other occurrences of $\#$. In the latter case the nonempty y must end with $\#$, implying that the right side has more $\#$s than the left side. The same implication holds in the case where γ_{i_j} is a letter of Δ. The contradiction proves our claim and, hence we have reached the subgoal.

We have now completed our first task. The second task still remains: we have to eliminate the prefix $\bar{u}\#$ and the suffix $\bar{v}\#$ in order to obtain the instance PCP that we are looking for.

Starting from $\bar{u}\#$, $\bar{v}\#$ and the pairs

$$(\gamma_i, \delta_i), \quad i = 1,\ldots,n + \text{card}(\Delta) + 1$$

we construct the pairs

$$(\gamma'_i, \delta'_i), \quad i = 1,\ldots,n + \text{card}(\Delta) + 3,$$

as follows. The words γ'_i and δ'_i will be over the alphabet $\Delta \cup \{\#, B\}$, where B is a new letter which acts as a separator. For a nonempty word x over $\Delta \cup \{\#\}$, we denote by ^Bx (resp. x^B) the word obtained from x by writing B before (resp. after) each letter. (Thus, $^Baba = BaBbBa$ and $aba^B = aBbBaB$.)

Now define

$$\gamma'_1 = BB, \quad \delta'_1 = B^B(\bar{u}\#), \quad \gamma'_2 = (\bar{v}\#)^BB, \quad \delta'_2 = BB,$$

$$\gamma'_i = \gamma^B_{i-2}, \quad \delta'_i =^B\!\delta_{i-2}, \quad i = 3,\ldots,n + \text{card}(\Delta) + 3.$$

Let g_2 and h_2 be the morphisms defined by

$$g_2(i) = \gamma'_i, \quad h_2(i) = \delta'_i, \quad i = 1,\ldots,n + \text{card}(\Delta) + 3.$$

We claim that $(*)_2$ holds for some w iff

$(*)_3$ $$h_2(w') = g_2(w')$$

holds for some nonempty w'. Consequently, by the subgoal already reached, v is in $L(S)$ iff $PCP = (g_2, h_2)$ possesses a solution. This completes the reduction, and Theorem 2.1 follows. In comparison with the instance determined by the pairs (α_i, β_i) of words over Δ, PCP has $\text{card}(\Delta) + 3$ more pairs and they are over the alphabet $\Delta \cup \{\#, B\}$.

Our claim concerning the equivalence between $(*)_2$ and $(*)_3$ remains to be shown. If $w = i_1 \ldots i_k$, $k \geq 0$, is a solution of $(*)_2$, then $w' = 1(i_1 + 2) \ldots (i_k + 2)2$ is a solution of $(*)_3$. This follows because if w renders both sides of $(*)_2$ equal to $y_1 \ldots y_l$, $l \geq 1$, where y_i are letters, then w' renders both sides of $(*)_3$ equal to

$$BBy_1 By_2 \ldots By_l BB .$$

Conversely, assume that $(*)_3$ holds for some nonempty w'. Without loss of generality we assume that $(*)_3$ holds for no proper prefix of w'. (This means that w' is a primitive or, more accurately, a suffix primitive solution of PCP.) Denote $w' = i_1 i_2 \ldots i_k$. We must have $i_1 = 1$ because, for all indices $i \neq 1$, γ_i' and δ_i' begin with a different letter. The same argument concerning the last letters of γ_i' and β_i' shows that $i_k = 2$. If $k = 2$, then $\bar{u}\# = \bar{v}\#$ and, consequently, $w = \lambda$ satisfies $(*)_2$. If $k > 2$ and none of the indices $i_2, \ldots, i_{k-1}$ is 1 or 2, then we see, exactly as before, that

$$w = (i_2 - 2) \ldots (i_{k-1} - 2)$$

satisfies $(*)_2$. Let $k > 2$ and let i_j be the first of the indices $i_2, \ldots, i_{k-1}$, that is, 1 or 2. If $i_j = 1$ then the second occurrence of BB on the right side of $(*)_3$ is immediately followed by another B, whereas on the left side it is immediately followed by a letter different from B, which is a contradiction. But if $i_j = 2$ then the proper prefix $i_1 \ldots i_j$ of w' also satisfies $(*)_3$, which contradicts our assumption concerning w'. Hence, we have established our claim concerning the equivalence of $(*)_2$ and $(*)_3$.

□

A reader familiar with other proofs of Theorem 2.1 will notice that a reduction to normal systems is much simpler. Indeed, a proof based on a reduction to the halting problem or to the membership problem of type 0 languages is considerably longer if presented in detail. Technical difficulties arise, for instance, from the necessity to take care of the correct parity of the length of computations or derivations, as well as to adding idle steps to derivations or computations.

2.2 Analyzing PCP further

Somewhat more can be said. According to Theorem 2.1, no algorithm can settle all instances of the Post Correspondence Problem. On the other hand, an algorithm

Analyzing PCP further 57

can be constructed for some classes of instances. We begin with an easy example.

Example 2.6. Consider all instances $PCP = (g, h)$, where the range alphabet Δ of g and h consists of one letter a only. Thus, the words $g(w)$ and $h(w)$ are powers of a. If we have $g(i) = h(i)$ for some i in the domain alphabet Σ, then PCP possesses a solution. Assume that $g(i) \neq h(i)$ for all i. Then, clearly, PCP has a solution exactly in case there are letters i and j of Σ such that $|g(i)| < |h(i)|$ and $|g(j)| > |h(j)|$. Hence, the special case $\Delta = \{a\}$ of the Post Correspondence Problem is decidable.

□

We can also take subclasses of all instances and still retain undecidability. First consider again an easy example.

Example 2.7. We now restrict our attention to all instances where the cardinality of the range alphabet is two: $\Delta = \{a, b\}$. We show that there can be no algorithm to settle all such instances. Assume the contrary: there is such an algorithm A. Consider an arbitrary instance $PCP' = (g, h)$ with the range alphabet

$$\Delta' = \{a_1, \ldots, a_k\}.$$

Let $f : \Delta'^* \to \Delta^*$ be the morphism defined by

$$f(a_i) = ba^i, \quad i = 1, \ldots, k.$$

Thus, we encode the letters of Δ' using the two letters of Δ. The morphism f is injective, which means that every word over Δ can be decoded in at most one way. Consequently, the instance $PCP = (fg, fh)$ has the same solutions as PCP'. Since the range alphabet of PCP is Δ, the algorithm A can be applied to solve PCP. We thus have an algorithm to solve all instances of the Post Correspondence Problem, a contradiction.

□

An undecidability result is strengthened if the scope of instances of our undecidable problem is restricted but undecidability is still retained. A decidability result is similarly strengthened by enlarging the scope of instances. By both of these methods, one approaches the borderline between decidability and undecidability. According to Examples 2.6 and 2.7, the borderline for the Post Correspondence Problem lies between 1 and 2, as far as the cardinality of the range alphabet is concerned. However, this result does not say much. It only indicates that two letters suffice to encode an arbitrary number of letters in such a way that the resulting morphism is injective. From other points of view, the gap between the cardinalities 1 and 2 is still huge.

In this section we will explore further the borderline between decidability and undecidability in the Post Correspondence Problem. We will also discuss some slightly different representations of the problem. The discussion illustrates the contents of Section 2.1 further.

Matters become much more complicated if we consider the cardinality n of the domain alphabet Σ. Recall that n is the number of words in each of the two lists if we follow the original formulation of the Post Correspondence Problem in Section 2.1. For a fixed $n \geq 1$, we can thus speak of the Post Correspondence Problem of *length n*. If $n = 1$, then a solution exists iff the single words in the two lists are identical.

What about the length 2? This case also seems easy at a first glance. An instance is determined by two pairs (α_1, α_2) and (β_1, β_2). After ruling out some obvious cases, we are left with a few possibilities which should lead to decidability. However, a more careful analysis shows that, no matter how we started our classification, the number of subcases tends to grow at each stage. Indeed, the decidability of the Post Correspondence Problem of length 2 requires a rather sophisticated argument. The case of length 3 seems very difficult to settle.

The statement of Theorem 2.1 does not as such tell us that we must have undecidability for a fixed length m. However, this follows from the proof of Theorem 2.1. Following the notation of the proof, we assume that the normal system S generates a nonrecursive set S. We may assume that the cardinality of Δ equals 2 because we may always apply the encoding technique of Example 2.7. The proof shows that the Post Correspondence Problem of length $n + 5$ is undecidable, where n is the number of rewriting rules in S. Since the boundary marker $\#$ can also be encoded in terms of the two letters of Δ, we obtain the bound $n + 4$ as well here. We state this result in the following theorem.

Theorem 2.2. *If a nonrecursive set is generated by a normal system with n rules, then the Post Correspondence Problem of length $n + 4$ is undecidable.*

A more detailed analysis along the lines of Theorem 2.2 shows that the Post Correspondence Problem of length 9 is undecidable. For more details concerning the next theorem, we refer the reader to [Sal 2].

Theorem 2.3. *The Post Correspondence Problem of length 2 is decidable. The Post Correspondence Problem of length 9 is undecidable.*

Theorem 2.3 leaves open the lengths from 3 to 8. Theorem 2.2 gives an interconnection between (i) the size of the smallest normal system generating a nonrecursive set and (ii) the length for which the Post Correspondence Problem is undecidable. It is very likely that the bound $n + 4$ can be improved further. One may also consider other devices for generating nonrecursive sets, such as Turing machines.

There are other angles from which the borderline between decidability and undecidability can be investigated. For instance, one may try to impose additional conditions on the morphisms g and h and still retain undecidability. By definition, a morphism $g : \Sigma^* \to \Delta^*$ is a *biprefix code* if, for any distinct letters i and j of Σ, $g(i)$ is neither a prefix nor a suffix of $g(j)$.

Theorem 2.4. *The Post Correspondence Problem is undecidable for instances (g, h), where g and h are biprefix codes.*

The proof of Theorem 2.4 is omitted. However, it is fairly easy to obtain an intuitive idea of the proof. The pairs $(g(i), h(i))$ are viewed as instructions of a deterministic Turing machine. Thus, $g(i)$ indicates the state and the scanned letter, and $h(i)$ indicates what happens. The pair (state, letter) is not the same for two distinct values of i. This is the reason why g can be chosen to be a biprefix code. As regards h, the matter is somewhat more complicated: one has to modify the behaviour of the Turing machine.

Theorem 2.4 deals with the sets

$$\{g(i)|i \in \Sigma\} \text{ and } \{h(i)|i \in \Sigma\}.$$

It does not say anything about the prefix and suffix relations between $g(i)$ and $h(i)$. Of course, in order to obtain a solution, there must be values i such that $g(i)$ is a prefix of $h(i)$, or vice versa.

We also wish to point out that there is no known decidable counterpart of Theorem 2.4. There is no known "reasonable" class of morphisms such that the Post Correspondence Problem is decidable for this class. However, our next theorem contains a result in this direction. Decidability is obtained if one of the morphisms belongs to a very small class while the other is general, in contrast to the morphisms in Theorem 2.4.

By definition, a morphism $h : \Sigma^* \to \Delta^*$ is *periodic* if there is a nonempty word w over Δ (called the *period* of h) such that $h(a)$ is a positive power of w for every letter a of Σ.

Example 2.8. Consider first the simple case where both the morphisms g and h are periodic. Let the g- and h-values be powers of u and w respectively. If neither one of the words u and w is a prefix of the other, the instance (g, h) clearly has no solution. The case $u = w$ is handled exactly as in Example 2.6. Thus, we are left with the case where u is a proper prefix of w:

$$w = uu_1 \text{ with } u_1 \neq \lambda.$$

The instance (g, h) can have a solution only if $u^i = w^j$ holds for some i and j. Hence,

(*) $$u^{i-1} = u_1(uu_1)^{j-1}.$$

If $j = 1$ then w is a power of u, which means that we are back in Example 2.6. Assume that w is not a power of u. Then one of the words u and u_1 is a proper prefix of the other. Assume that

$$u = u_1 u_2, \ u_2 \neq \lambda.$$

Taking from both sides of (*) the prefix of length $|u| + |u_1|$, we obtain $u_1 u_2 u_1 = u_1 u_1 u_2$ and, hence, $u_1 u_2 = u_2 u_1$. This shows (see Theorem 1.8) that u_1 and u_2 are powers of the same word. Hence, u and w are powers of this word, and we are again back in Example 2.6.

The case remains where u is a proper prefix of u_1. We write u_1 in the form $u_1 = u^k u_2$, where $k \geq 1$ and $1 \leq |u_2| < |u|$. From (*) we see that u_2 is also a proper suffix of u, i.e., $u = u_3 u_2$, for some nonempty u_3. We now divide (*) from the left by u^k, yielding

$$u^{i-k-1} = u_2(u^{k+1}u_2)^{j-1}.$$

The equation $u_3 u_2 = u_2 u_3$ now results by considering the prefix of length $|u_2 u_3|$ of both sides. But this again implies that u_2 and u_3, and hence also u and w, are powers of the same word.

We have shown that the Post Correspondence Problem is decidable if both g and h are periodic. Moreover, we have shown that the instance (g, h) can have a solution only if g and h have the same minimal period.

□

The argument given above is based on the combinatorics of words. The proof of the following stronger result uses facts from language theory.

Theorem 2.5. The Post Correspondence Problem is decidable for instances (g, h), where one of g and h is periodic.

Proof. Assume that h is periodic with period w. Consider the languages

$$L_1 = g^{-1}(w^*) \text{ and } L_2 = \{x \in \Sigma^+ \mid |g(x)| = |h(x)|\}.$$

The intersection $L = L_1 \cap L_2$ consists of all solutions of the instance (g, h) of the Post Correspondence Problem. Indeed, $x \in L$ implies that $g(x)$ is a power of w and that $|g(x)| = |h(x)|$ and, consequently, that $g(x) = h(x)$ because $h(x)$ is always a power of w. Conversely, if x is a solution then x is in both L_1 and L_2. Thus, solving the Post Correspondence Problem amounts in this case to deciding the emptiness of L.

Recall that $\psi(L)$ denotes the set of Parikh vectors of words in L. Clearly, L is empty iff $\psi(L)$ is empty. Moreover,

$$\psi(L) = \psi(L_1) \cap \psi(L_2).$$

This equation (which is not valid in general) is valid because in L_2 the order of letters is immaterial: if x is in L_2, then any permutation of x is also in L_2.

The emptiness of the intersection $\psi(L_1) \cap \psi(L_2)$ is decidable because of the following three facts. (i) The language L_1 is regular and, hence, $\psi(L_1)$ is effectively semilinear. (ii) Also, $\psi(L_2)$ is effectively semilinear because it consists of nonnegative solutions of a linear equation. (iii) The intersection of two semilinear sets is effectively semilinear. See Theorem 1.12.

□

Theorems 2.3 and 2.5 exhibit the only known nontrivial classes of instances for which the Post Correspondence Problem is decidable.

Let us look at some modifications. Let g and h be morphisms mapping Σ^* into Δ^*, and u_1, u_2, v_1, v_2 words over Δ. The sixtuple

$$GPCP = (g, h, u_1, u_2, v_1, v_2)$$

is called an *instance* of the *Generalized Post Correspondence Problem*. A nonempty word w over Σ such that

$$u_1 h(w) u_2 = v_1 g(w) v_2$$

is referred to as a *solution* of $GPCP$.

An algorithm for finding a solution of the generalized version can be applied for finding a solution of the original Post Correspondence Problem: one simply chooses the words u_i and v_i to be empty. In this sense, the generalized version is "more difficult" than the original one. However, it only looks more difficult. The next theorem shows that the generalized version is, in fact, the original version in disguise. As before, we define the *length* of $GPCP$ to be the cardinality of Σ.

Theorem 2.6. For each instance $GPCP$ of length n, an instance PCP of length $n+2$ can be constructed such that PCP has a solution iff $GPCP$ has a solution.

Proof. Assume that

$$GPCP = (g, h, u_1, u_2, v_1, v_2),$$

and that g and h are determined by their values γ_i and δ_i, $i = 1, \ldots, n$. (Thus, for letters i of Σ, $g(i) = \gamma_i$ and $h(i) = \delta_i$. As before, we visualize γ_i and δ_i, $i = 1, \ldots, n$, to be the two lists.) We use the concept from the end of the proof of Theorem 2.1. We use also the same notation. For a nonempty word x, $^B x$ (resp. x^B) denotes the word obtained from x by writing B before (resp. after) each letter. By definition, $^B\lambda = \lambda^B = \lambda$.

The instance $PCP = (g', h')$ is defined as follows. Two new letters are added to the domain alphabet Σ of g and h, and the letter B to the range alphabet Δ. The morphisms g' and h' are determined by the values γ'_i and δ'_i:

$$\gamma'_1 = BBv_1^B,\ \delta'_1 = B^B u_1,\ \gamma'_2 = v_2^B B,\ \delta'_2 = {}^B u_2 BB$$

$$\gamma'_i = \gamma_{i-2}^B,\ \delta'_i = {}^B \delta_{i-2},\ i = 3, \ldots, n+2.$$

The argument is now similar to the one given in the proof of Theorem 2.1. Assume that $w = i_1 \ldots i_k$, $k \geq 1$, is a solution of $GPCP$, i. e.,

$$u_1 h(w) u_2 = u_1 \delta_{i_1} \ldots \delta_{i_k} u_2 = v_1 \gamma_{i_1} \ldots \gamma_{i_k} v_2 = v_1 g(w) v_2.$$

Then

$$w' = 1(i_1 + 2) \ldots (i_k + 2) 2$$

is a solution of PCP because

$$h(w') = B^B u_1^B \delta_{i_1} \ldots {}^B \delta_{i_k}^B u_2 BB = BBv_1^B \gamma_{i_1}^B \ldots \gamma_{i_k}^B v_2^B B = g(w').$$

Indeed, both $h(w')$ and $g(w')$ result from the word

$$u_1 h(w) u_2 = v_1 g(w) v_2$$

by separating the letters with Bs and adding BB to the beginning and the end.

Conversely, assume that w' is a solution of PCP such that no proper prefix of w' is a solution. We conclude, exactly as in the proof of Theorem 2.1, that w' must begin with the letter 1 and end with the letter 2 and that the intermediate letters yield a solution of $GPCP$.

□

The given proof actually shows more than is claimed in the theorem. It also provides a method of transforming an arbitrary solution of $GPCP$ to a solution of PCP, and vice versa. If we restrict our attention to some specific class of morphisms, the proof tells us to what extent this class is preserved in the transition from $GPCP$ to PCP.

In the *modified version* of the Post Correspondence Problem we are interested only in solutions that begin and/or end in a prescribed fashion. For instance, we might only look for solutions that begin with the index 2 and end with the sequence 313. We now give the formal definition.

Again let g and h be morphisms mapping Σ^* into Δ^*, and let w_1 and w_2 be words over Σ. Then the quadruple

$$MPCP = (g, h, w_1, w_2)$$

is termed an *instance* of the *Modified Post Correspondence Problem*. A nonempty word w over Σ such that $g(w_1 w w_2) = h(w_1 w w_2)$ is referred to as a *solution* of $MPCP$.

The decidability of the Modified Post Correspondence Problem can be viewed from different angles. There can be no algorithm for solving all instances because instances of the Post Correspondence Problem are obtained as special cases by choosing $w_1 = w_2 = \lambda$. On the other hand, in many applications w_1 and w_2 are fixed. For instance, we might be interested only in solutions that begin with the index 1. The problem always remains undecidable, as can be seen in the next theorem.

Theorem 2.7. Let the words w_1 and w_2 be fixed. Then there is no algorithm for solving all instances (g, h, w_1, w_2) of the Modified Post Correspondence Problem.

Proof. The proof of Theorem 2.1 establishes the claim for $w_1 = 1$ and $w_2 = 2$. More specifically, it was shown that there is no algorithm for solving all instances (g, h) of the Post Correspondence Problem such that one of $g(i)$ and $h(i)$ is a prefix of the other (resp. one of $g(j)$ and $h(j)$ is a suffix of the other) exactly in the case $i = 1$ (resp. $j = 2$). As usual, let the alphabets be Σ and Δ.

Now consider arbitrary w_1 and w_2. We show how an algorithm A for solving all instances (g, h, w_1, w_2) can be applied to solve all instances (g, h) described in the preceding paragraph, which is a contradiction.

Analyzing PCP further

Let such an instance (g, h) be fixed, where Σ has n letters. Assume that $|w_1| = k$, $|w_2| = l$, and that $b_1, \ldots, b_r$ are all the distinct letters occurring in $w_1 w_2$. We assume first that both w_1 and w_2 are nonempty, and that the first letter of w_1 is not the same as the last letter of w_2. We choose the notation in such a way that b_1 is the first letter of w_1, and b_2 is the last letter of w_2.

We add the letters b_i as new letters to both the alphabets Σ and Δ, as well as the letter a as a new letter to the alphabet Δ. The morphisms g' and h', operating with the augmented alphabets, are defined in terms of g and h by the following table:

x	b_1	b_2	$b_j, j > 2$	1	2	$3 \leq i \leq n$
$g'(x)$	a	a	a	$w_1' g(1)$	$g(2) w_2'$	$g(i)$
$h'(x)$	$a^k b_1$	$b_2 a^l$	b_j	$h(1)$	$h(2)$	$h(i)$

Here w_1' (resp. w_2') is obtained from w_1 (resp. w_2) by replacing all occurrences of b_1 and b_2 with $a^k b_1$ and $b_2 a^l$, except that the first letter of w_1 (resp. the last letter of w_2) is left intact.

A straightforward analysis shows that the instance $MPCP = (g', h', w_1, w_2)$ has a solution iff the instance $PCP = (g, h)$ has a solution. Indeed, the solutions of $MPCP$ coincide with words of the form

$$u_1 w_2 w_1 u_2 w_2 \ldots w_1 u_s,$$

where $s \geq 1$ and each u_i is a primitive solution of PCP. On the other hand, nonprimitive solutions of PCP are not solutions of $MPCP$.

Let us consider an example in order to illustrate how things work. Assume that

$$w_1 = b_1 b_2 b_1 \text{ and } w_2 = b_3 b_1 b_3 b_2.$$

Then the morphisms g' and h' are defined by the following table:

x	b_1	b_2	b_3	1	2	$3 \leq i \leq n$
$g'(x)$	a	a	a	$b_1 b_2 a^7 b_1 g(1)$	$g(2) b_3 a^3 b_1 b_3 b_2$	$g(i)$
$h'(x)$	$a^3 b_1$	$b_2 a^4$	b_3	$h(1)$	$h(2)$	$h(i)$

If a nonempty word u satisfies $g'(u) = h'(u)$, then u must begin with b_1 because no other letter gives two candidates, one of which is a prefix of the other. Only the prefix $b_1 b_2 b_1$ leads to an eventual elimination of the bs. (Just to eliminate a^3, we could start with any three b-letters.) If now $g(w) = h(w) = z$ and 1 and 2 occur in w only once (which is equivalent to w being a primitive solution), then

$$g'(b_1b_2b_1w) = a^3b_1b_2a^7b_1zb_3a^3b_1b_3b_2,$$
$$h'(b_1b_2b_1w) = a^3b_1b_2a^4a^3b_1z.$$

Hence,
$$g'(b_1b_2b_1wb_3b_1b_3b_2) = h'(b_1b_2b_1wb_3b_1b_3b_2)$$
$$= a^3b_1b_2a^7b_1zb_3a^3b_1b_3b_2a^4.$$

Briefly, the definition of g' and h' "forces" the solutions of $MPCP$ and PCP to have the stated interdependence.

Essentially the same construction works if one of the words w_1 and w_2 is empty. Then either b_1 and the prefix w_1' of $g'(1)$, or else b_2 and the suffix w_2' of $g'(2)$ are omitted. Of course, there is nothing to prove if both w_1 and w_2 are empty.

There still remains the case where the first letter b_1 of w_1 equals the last letter of w_2. In this case we have to deposit the as in $h'(b_1)$ in a symmetric fashion. Denote

$$m = \max(k,l), \; k_1 = m - k, \; l_1 = m - l.$$

The morphisms g' and h' are now defined by the following table:

x	b_1	$b_j, j>1$	1	2	$3 \leq i \leq n$
$g'(x)$	a	a	$a^{k_1}w_1''g(1)$	$g(2)w_2''a^{l_1}$	$g(i)$
$h'(x)$	$a^m b_1 a^m$	b_j	$h(1)$	$h(2)$	$h(i)$

Here w_1'' and w_2'' are obtained from w_1 and w_2 by replacing every occurrence of b_1 by $a^m b_1 a^m$, except that the first occurrence in w_1 is replaced by $b_1 a^m$ and the last occurrence in w_2 by $a^m b_1$.

Modify our previous example in such a way that

$$w_1 = b_1b_2b_1 \text{ and } w_2 = b_3b_1b_3b_1.$$

Thus, b_2 now assumes the role of a "normal" letter. The definition of g' and h' now reads as follows:

x	b_1	b_2	b_3	1	2	$3 \leq i \leq n$
$g'(x)$	a	a	a	$ab_1a^4b_2a^4b_1a^4g(1)$	$g(2)b_3a^4b_1a^4b_3a^4b_1$	$g(i)$
$h'(x)$	$a^4b_1a^4$	b_2	b_3	$h(1)$	$h(2)$	$h(i)$

If $g(u) = h(u) = z$ and 1 and 2 occur in u only once, then

$$g'(w_1 u w_2) = h'(w_1 u w_2) = a^4 b_1 a^4 b_2 a^4 b_1 a^4 z b_3 a^4 b_1 a^4 b_3 a^4 b_1 a^4 \ .$$

□

The interrelation between $GPCP$ and $MPCP$ is obvious:

$$(g, h, w_1, w_2) \quad \text{and} \quad (g, h, h(w_1), h(w_2), g(w_1), g(w_2))$$

have the same solutions. Consequently, a result analogous to Theorem 2.7 can be immediately obtained for $GPCP$s.

2.3 Equality sets and test sets

What has been said above can be looked at from a different point of view. The solutions of an instance PCP constitute a language. It shows the generality of the Post Correspondence Problem that the languages obtained in this fashion represent, in the sense made precise below, all recursively enumerable languages.

As before, let g and h be morphisms mapping Σ^* into Δ^*. The *equality set* or *equality language* between g and h is defined by

$$E(g, h) = \{w \in \Sigma^+ | g(w) = h(w)\} \ .$$

Thus, the empty word λ is not in $E(g, h)$, although $g(\lambda) = h(\lambda)$. This reflects the fact that λ is not considered to be a solution of the instance $PCP = (g, h)$. Consequently, it follows by the definitions that the equality set $E(g, h)$ is nonempty iff the instance $PCP = (g, h)$ possesses a solution.

Example 2.9. We consider some equality sets by first returning to Example 2.4. Thus, define the morphisms by

$$g(1) = a^2 b, \ \ g(2) = a^2, \ \ h(1) = a, \ \ h(2) = ba^2 \ .$$

An attempt to construct a word w in $E(g, h)$ leads us to the following table. At every step the addition of a new letter to the end of w is *forced* in the sense that it is the only possibility of preserving the necessary condition of one of the words $g(w)$ and $h(w)$ being a prefix of the other:

w	1	11	112	1122	11221
$g(w)$	$a^2 b$	$a^2 b a^2 b$	$a^2 b a^2 b a^2$	$a^2 b a^2 b a^4$	$a^2 b a^2 b a^6 b$
$h(w)$	a	a^2	$a^2 b a^2$	$a^2 b a^2 b a^2$	$a^2 b a^2 b a^3$
to catch	ab	$ba^2 b$	ba^2	a^2	$a^3 b$

Here "to catch" indicates the missing suffix of the shorter of the words $g(w)$ and $h(w)$, in this case always $h(w)$. In order to catch a^3b, the next three letters have to be 111.

For $w = 11221111$, there is the word $u = ba^2b(a^2b)^{2^2-2}$ to catch. But notice that, for $w = 11$, there is the prefix ba^2b of u to catch. It is easy to see that, after a while, we have the word $ba^2b(a^2b)^{2^3-2}$ to catch, and so forth. Consequently, we never reach a word w with $g(w) = h(w)$, which means that $E(g,h)$ is empty. However, arbitrarily long words w can be constructed such that $h(w)$ is a prefix of $g(w)$. In fact, our argument above shows that all prefixes of the infinite word

$$1122111122221^8 2^8 1^{16} 2^{16} \ldots ,$$

and no other words, have this property.

We next change the value $g(2)$ and obtain a new pair (g, h):

$$g(1) = a^2 b, \quad g(2) = a, \quad h(1) = a, \quad h(2) = ba^2.$$

As before, every step is forced in the construction of a candidate word w. We now obtain the table

w	1	11	112	1122
$g(w)$	a^2b	a^2ba^2b	a^2ba^2ba	$a^2ba^2ba^2$
$h(w)$	a	a^2	a^2ba^2	$a^2ba^2ba^2$
to catch	ab	ba^2b	ba	$---$

Thus, we now reach a word w with $g(w) = h(w)$. Since after this word we may start all over again and every step is forced, we conclude that

$$E(g, h) = \{(1122)^i | i \geq 1\}.$$

This equality set is very low in all language hierarchies. It is a regular language that is accepted by a very simple automaton. However, equality sets can be very complex. The reason for this will become apparent in Theorem 2.8. We now exhibit a further modification of our example which gives rise to an equality set that is not a context-free language.

We add a few more letters to the alphabets and define the pair (g, h) by

i	1	2	3	4	5
$g(i)$	a^2b	a^2	c	c	c
$h(i)$	a	ba^2	ba^2	bc	c^2

Equality sets and test sets 67

Observe first that, for the argument values 1 and 2, the morphisms coincide with our first variant. The remaining argument values make it possible for h to catch up words of the form $ba^2b(a^2b)^{2^{j+1}-2}$, $j \geq 0$. Words of this form occur exactly in case

(*) $$w = 1^2 2^2 1^4 2^4 1^8 2^8 \ldots 1^{2^{j+1}}.$$

For such a w, catching up is described as follows:

	w	$w3^{2^{j+1}-1}$	$w3^{2^{j+1}-1}45^{2^{j+1}-1}$
to catch	$ba^2b(a^2b)^{2^{j+1}-2}$	$bc^{2^{j+1}-1}$	---

We can now prove that $E(g,h) = L^+$, where

$$L = \{1^2 2^2 1^4 2^4 \ldots 1^{2^{j+1}} 3^{2^{j+1}-1} 45^{2^{j+1}-1} | j \geq 0\}.$$

We have seen that words in L^+ are in $E(g,h)$. To establish the reverse inclusion, it suffices to analyze the construction of a candidate word in $E(g,h)$. Now we can begin with $i = 5$ but then it is easy to see that a word in $E(g,h)$ never results. Thus, we have to begin with $i = 1$. Then, as seen above, every step is forced and prefixes of words of the form (*) result, provided we use only the argument values 1 and 2. For words (*), we may start the terminating path beginning with 3. If the terminating path is started but then interrupted with values 1 or 2, the first c can never be eliminated. The same thing happens if one tries to initiate a terminating path for a word not of the form (*). This shows that $E(g,h) = L^+$.

Finally, consider the historically interesting pair (g,h) defined by Post (see Example 2.5):

$$g(1) = b^2, \quad g(2) = ab, \quad g(3) = b,$$
$$h(1) = b, \quad h(2) = ba, \quad h(3) = b^2.$$

When reading words over the alphabet $\{1,2,3\}$, we apply a *counter* as follows. Initially, the counter is set at 0. When a letter is read, 1, 0 or -1 is added to the counter value, depending on whether the letter is 1, 2 or 3. (Thus, the counter value of a word w indicates how many more bs there are in $g(w)$ than in $h(w)$.) For instance, the words

$$13122213, \quad 113113233, \quad 3112231312133$$

have counter values $+1$, 0 and 0 respectively.

Let L be the language over $\{1,2,3\}$ consisting of all nonempty words such that (i) the counter value of w equals 0, and (ii) each occurrence of the letter 2 in w, if any, is preceded by a prefix with the counter value $+1$. For instance, the first of the words mentioned above fails to satisfy (i), the second fails to satisfy (ii), whereas the third satisfies both (i) and (ii). A reader who has worked through the much more complicated example above will have no difficulties in showing that $L = E(g,h)$. It can be added that L is context-free but not regular and that $L = L^+$.

□

While equality sets are clearly recursive, the next theorem shows that their expressive power is still quite remarkable. In fact, every recursively enumerable language is obtained from an equality language by first restricting attention to words of a particular form and then erasing all occurrences of some letters from these words.

More specifically, let L be a language over Σ. Let b and e be letters of Σ (possibly $b = e$), and Σ_1 and Δ subalphabets of Σ. We say that L' over Δ results from L by *ordering* and *erasing* if

$$L' = \{w_\Delta | b w_{\Sigma_1} w_\Delta e \in L,\ w_{\Sigma_1} \in \Sigma_1^*,\ w_\Delta \in \Delta^*\}.$$

Equivalently,

$$L' = f_\Delta(L \cap b\Sigma_1^* \Delta^* e),$$

where $f_\Delta : \Sigma^* \to \Delta^*$ is the morphism defined by

$$f_\Delta(a) = \begin{cases} a & \text{for } a \in \Delta, \\ \lambda & \text{for } a \in \Sigma - \Delta. \end{cases}$$

Theorem 2.8. *Every recursively enumerable language results from some equality set by ordering and erasing.*

Proof. We apply a part of the proof of Theorem 2.1. We choose the normal system S considered there to generate the *mirror image* of the given recursively enumerable language L' over the alphabet Δ:

$$L(S) = \overline{L'}.$$

Consider also the normal system S''' from the proof of Theorem 2.1. Thus, we have, as before, the equivalence

$$\bar{v}\# \text{ is in } L(S''') \text{ iff } v \text{ is in } L(S).$$

This equivalence can now be written

$$v\# \text{ is in } L(S''') \text{ iff } v \text{ is in } L'.$$

Let $u\#$ be the axiom of S''' and

$$\alpha_i X \to X \beta_i, \quad i = 1, \ldots, n,$$

be the rules. (Here we have simplified the previous notation.) The system S''' has been constructed in such a way that $v\#$ is in $L(S''')$ exactly in case there is a sequence of indices $i_1, \ldots, i_k$ such that

(*) $$u\# \beta_{i_1} \beta_{i_2} \ldots \beta_{i_k} = \alpha_{i_1} \alpha_{i_2} \ldots \alpha_{i_k} v\#.$$

Denote by Δ_1 the alphabet of S'''. Thus, Δ_1 consists of the letters of Δ, # and of the other letters that possibly occur in α_i, β_i and u. Denote, further,

$$\Sigma = \Sigma_1 \cup \Delta \cup \{b, e\} \text{ with } \Sigma_1 = \{1, \ldots, n\}.$$

We now define two morphisms g and h mapping Σ^* into Δ_1^*.

$$\begin{aligned} g(i) &= \alpha_i, & h(i) &= \beta_i, & \text{for } i \in \Sigma_1, \\ g(a) &= a, & h(a) &= \lambda, & \text{for } a \in \Delta, \\ g(b) &= \lambda, & h(b) &= u\#, \\ g(e) &= \#, & h(e) &= \lambda. \end{aligned}$$

We now claim that L' results from $E(g, h)$ by ordering and erasing. More specifically,

(**) $$L' = f_\Delta(E(g, h) \cap b\Sigma_1^* \Delta^* e),$$

where the morphism f_Δ is defined before the statement of the theorem.

Indeed, the claim is an almost immediate consequence of the definitions. Assume first that v is in L'. This implies, as shown above, that $v\#$ is in $L(S''')$. This, in turn, implies that (*) is satisfied for some sequence $i_1, \ldots, i_k$. (Here we may have $k = 0$, meaning that $i_1 \ldots i_k = \lambda$.) Denoting $w = bi_1 \ldots i_k ve$, we see that $h(w)$ equals the left side of (*), whereas $g(w)$ equals the right side of (*). Consequently, w is in $E(g, h)$ and v in the language on the right side of (**). (Observe that the argument also holds for $v = \lambda$.)

Conversely, assume that v is in the language on the right side of (**). Hence, for some w_1 in Σ_1^*,

$$h(bw_1 ve) = g(bw_1 ve).$$

This means that (*) holds for some sequence of indices $i_1, \ldots, i_k$. Consequently, $v\#$ is in $L(S''')$. (This conclusion also holds if we are dealing with the empty sequence: $k = 0$.) But $v\#$ being in $L(S''')$ is equivalent to v being in L'. Hence, our claim follows. □

In comparison with other proofs of results similar to Theorem 2.8 the above approach, where material from Section 2.1 is used, is much simpler.

The idea in the preceding proof can be modified to yield another representation of recursively enumerable languages in terms of two morphisms. Instead of the equality set, we now use another construction in which we cut off the prefix $g(w)$ from $h(w)$. For three words u, v and w, we write the equation $w = uv$ in the alternative form $u^{-1}w = v$. (More specifically, the descriptive notation u^{-1} is used for the operation of cutting off the prefix u. If u is not a prefix of w, $u^{-1}w$ is undefined.)

Theorem 2.9. Every recursively enumerable language L' over Δ can be expressed in the form

$$L' = \{(g(w))^{-1} h(w) | w \in b\Sigma_1^* \Delta^* e\},$$

where g and h are morphisms, Σ_1 is an alphabet and b and e are letters.

Proof. We use the notations from the proof of Theorem 2.8 except that there is a slight difference in the definition of the morphism h: we now define $h(a) = a$ for $a \in \Delta$, instead of the previous $h(a) = \lambda$.

As before, v being in L' is equivalent to the existence of a sequence $i_1, \ldots, i_k$ such that (*) holds. But now, for $w = bi_1 \ldots i_k ve$, (*) implies the equation

$$(g(w))^{-1}h(w) = v .$$

Conversely, assume that this equation holds for some w in $b\Sigma_1^* ve$. We write $w = bi_1 \ldots i_k ve$. (Possibly $k = 0$.) Clearly,

$$g(w) = \alpha_{i_1} \ldots \alpha_{i_k} v\# \text{ and } h(w) = u\#\beta_{i_1} \ldots \beta_{i_k} v .$$

Hence, our assumption implies that (*) holds and, consequently, v is in L'.
□

We now look at equality sets from an entirely different angle. We will establish a result that is very important in language theory and also illustrates the fact that a set may possess a finite subset with certain properties, although we are not, even in principle, able to find the subset.

Two morphisms g and h are *equivalent* on a language $L \subseteq \Sigma^*$ if $g(w) = h(w)$ holds for every w in L, or, equivalently, if

$$L \subseteq E(g, h) \cup \{\lambda\} .$$

A *finite* set $F \subseteq L$ is a *test set* for L if whenever two morphisms g and h are equivalent on F, they are also equivalent on L. Thus, when considering test sets, we keep the language L fixed but let the morphisms g and h vary.

Example 2.10. We claim that $F = \{ab, a^2b^2\}$ is a test set for the language $L = \{a^n b^n | n \geq 1\}$. Let g and h be arbitrary morphisms equivalent on F. Denote $g(a) = x$, $g(b) = y$, $h(a) = x_1$, $h(b) = y_1$. Consequently,

(*) $$xy = x_1 y_1 \text{ and } xxyy = x_1 x_1 y_1 y_1 .$$

To prove that g and h are equivalent on L, we assume without loss of generality that $x = x_1 z$, where z is nonempty. Consequently, $y_1 = zy$. We now obtain from the second equation in (*)

$$x_1 z x_1 z y y = x_1 x_1 z y z y ,$$

yielding $z x_1 z y = x_1 z y z$ and, further,

$$z x_1 = x_1 z \text{ and } zy = yz .$$

These commuting relations are possible only if each of the words z, x_1 and y is a power of the same word v. Consequently, x and y_1 are also powers of v. The first equation in (*) now implies that g and h are equivalent on L.

Our test set F consists of two words. The language L does not have any test set consisting of one word only. This is seen by the following argument. For each $n = 1, 2, \ldots$, consider the morphisms g_n and h_n defined by

$$g_n(a) = a, \quad g_n(b) = ba^n, \quad h_n(a) = a^n b, \quad h_n(b) = a.$$

Because $g_n(a^m b^m) = h_n(a^m b^m)$ iff $m = n$, the pair (g_m, h_m) shows that the set $\{a^m b^m\}$, $m = 1, 2, \ldots$, is not a test set for L. □

We now turn to more general questions concerning test sets. In some cases two morphisms g and h can be equivalent on a language only if they are identical. For instance, consider languages over the alphabet $\{a, b\}$. For a nonempty word w, denote by $R_{ab}(w)$ the ratio between the number of occurrences of a and that of b in w. (The ratio is considered to be infinite if there are no bs in w.) Assume that $F = \{u, w\}$, where

(∗) $$R_{ab}(u) \neq R_{ab}(w).$$

Consider morphisms g and h satisfying

(∗∗) $$g(u) = h(u) \text{ and } g(w) = h(w).$$

We claim that $g = h$. This is obvious if one of the ratios is 0 or ∞. If this is not the case and $g \neq h$, then

$$|g(a)| - |h(a)| = \alpha \text{ and } |h(b)| - |g(b)| = \beta,$$

for some positive integers α and β. (We restrict our attention to one of the two symmetric cases.) By (∗∗) we now obtain

$$m_1 \alpha - n_1 \beta = 0 = m_2 \alpha - n_2 \beta,$$

where m_1/n_1 and m_2/n_2 are the two ratios appearing in (∗), which contradicts (∗).

We have also shown that if a language L over $\{a, b\}$ contains two words u and w satisfying (∗), then $\{u, v\}$ is a test set for L.

Methods are known by which test sets can be constructed for languages of specific types, such as regular and context-free languages. For instance, consider a regular language L accepted by a finite deterministic automaton A. The subset of L, consisting of all words read by A through a sequence of state transitions where no state is met more than twice, constitutes a test set for L. This can be established using the following argument inductively.

Assume that each of the words t, tx and ty take A from the initial state to the same state. (In other words, both x and y cause a loop.) Assume, further, that tu is in L. Consequently, txu, tyu and $txyu$ are also in L. Let g and h be morphisms satisfying

$$g(tu) = h(tu), \quad g(txu) = h(txu), \quad g(tyu) = h(tyu).$$

We claim that $g(txyu) = h(txyu)$. Indeed, this is obvious if $g(t) = h(t)$. Otherwise, we assume without loss of generality that $h(t) = g(t)z$. The above equations imply, for some z_1 and z_2, the following five equations

$$g(u) = zh(u),$$

$$g(x) = zz_1 \text{ and } h(x) = z_1 z, \text{ or else } z = g(x)z_1 = z_1 h(x),$$

$$g(y) = zz_2 \text{ and } h(y) = z_2 z, \text{ or else } z = g(y)z_2 = z_2 h(y).$$

The claim $g(txyu) = h(txyu)$ is now easy to verify in each of the resulting cases.

Rather than dwelling further on languages of specific types, we investigate the general claim of the existence of a test set for an arbitrary language, not even necessarily recursively enumerable. (Recall that test sets are finite.) This claim was referred to for a long time as *Ehrenfeucht's Conjecture*, until the existence was finally settled in 1985. The construction of a test set cannot be effective, which is due to many known language-theoretic undecidability results.

We now begin the proof of Ehrenfeucht's Conjecture. Our proof proceeds via three subgoals and is self-contained. The essential mathematical tool, Hilbert's Basissatz, is established on the way.

Consider words w_1 and w_2 over an alphabet V, referred to as the alphabet of *variables*. A morphism $h : V^* \to \Sigma^*$ satisfying $h(w_1) = h(w_2)$ is called a *solution* of the *word equation*

$$w_1 = w_2.$$

Let S be at most a denumerably infinite system of word equations

$$w_1^{(i)} = w_2^{(i)}, \quad i = 1, 2, \ldots,$$

where all words are over the same alphabet of variables. Also, a morphism h is now called a *solution* of S if $h(w_1^{(i)}) = h(w_2^{(i)})$ holds for all i. Two systems of word equations are termed *equivalent* if they possess the same solutions. We are now ready to formulate our first subgoal.

Goal 1. *Every system of word equations possesses a finite equivalent subsystem.*

Rather than proving this statement, at this stage we only show that Ehrenfeucht's Conjecture is implied by Goal 1. Let L be an arbitrary language (not necessarily recursively enumerable) over Σ. We consider two "indexed versions" of the alphabet Σ,

$$\Sigma_i = \{a_i | a \in \Sigma\}, \quad i = 1, 2,$$

as well as two morphisms

$$\psi_i : \Sigma^* \to \Sigma_i^* \text{ defined by } \psi_i(a) = a_i.$$

For every word w in L, we consider the word equation

$$\psi_1(w) = \psi_2(w)$$

with the alphabet of variables $\Sigma_1 \cup \Sigma_2$. By Goal 1, the resulting system S of word equations possesses a finite equivalent subsystem S'. Let w_i, $i = 1, \ldots, k$, be all the words appearing on either the left or the right side of some equation in S'. (There is only a finite number of such words. In general, we are not able to find them. Not even the system S is effectively constructable.)

It now follows that $F = \{w_1, \ldots, w_k\}$ is a test set for L. Indeed, if two morphisms g_1 and g_2 are equivalent on F, then the morphism h defined by

$$h\psi_j(a) = g_j(a), \quad a \in \Sigma; \quad j = 1, 2$$

is a solution of the system S'. Since S' and S are equivalent, h is also a solution of the system S, that is,

$$h\psi_1(w) = h\psi_2(w)$$

for every w in L. But this shows that g_1 and g_2 are equivalent on L and, consequently, F is a test set for L.

So far we have only made a rather trivial reduction. Our next step concerns *arithmetization* and carries us much further.

Consider polynomials with integer coefficients and with variables taken from a fixed alphabet of variables

$$X = \{x_1, \ldots, x_r\}.$$

An r-tuple of integers $(u_1, \ldots, u_r)$ is called a *solution* of the system P of *polynomial equations*

$$p_i = 0, \quad i = 1, 2, \ldots,$$

if each polynomial p_i assumes the value 0 when each x_j is assigned the value u_j, for $j = 1, \ldots, r$. (The total number of the polynomials is at most denumerably infinite.) Again, two systems of polynomial equations are called *equivalent* if they possess the same solutions. We now formulate our next subgoal.

Goal 2. Every system of polynomial equations possesses a finite equivalent subsystem.

We show now that Goal 2 implies Goal 1. Hence, by our previous result, it suffices to prove the statement of Goal 2 in order to establish Ehrenfeucht's Conjecture.

Thus, assume that the statement of Goal 2 is true and consider an arbitrary system S of word equations with the alphabet V of variables. As before, we again consider two indexed versions V_1 and V_2 of V. We associate with the system S a system $P = P(S)$ of polynomial equations as follows.

The set of variables of P equals $V_1 \cup V_2$. Each equation $w = w'$ in S gives rise to two equations

$$t_1(w) - t_1(w') = 0 \text{ and } t_2(w) - t_2(w') = 0,$$

where the mappings t_1 and t_2 remain to be defined. Indeed, $t_2 : V^* \to V_2^*$ is the morphism mapping each letter v into its indexed version v_2. However, we view the values of t_2 as monomials with coefficient 1. Moreover, t_1 is the mapping of V^*

into the set of polynomials with all coefficients equal to 1 and with variables taken from $V_1 \cup V_2$, defined by the conditions

$$t_1(\lambda) = 0,$$
$$t_1(v) = v_1 \text{ for } v \in V,$$
$$t_1(w_1 w_2) = t_1(w_1) t_2(w_2) + t_1(w_2).$$

Here 0 stands for the zero polynomial. It is obvious that the value of t_1 does not depend on the way in which the catenation is associated when the last equation is applied. For instance,

$$t_1(uvuu) = u_1 v_2 u_2 u_2 + v_1 u_2 u_2 + u_1 u_2 + u_1,$$

and the same polynomial results, no matter how we choose w_1 and w_2 in the last of the defining equations.

Let P' be a finite equivalent subsystem of the system P. For every polynomial equation in P', there is a word equation in S such that the polynomial equation results from the word equation. Let S' be the finite subsystem of S obtained in this fashion. (When we go back from S' to polynomial equations, we obtain a system P'' containing P' and possibly some other equations. This is due to the fact that every equation in S gives rise to two equations in P and possibly only one of them is in P'.) We claim that S' is equivalent to S.

Assume the contrary: a morphism $h : V^* \to \Sigma^*$ is a solution of S', whereas $h(y) \neq h(y')$, for some equation $y = y'$ in $S - S'$.

Let $\Sigma = \{a_0, \ldots, a_{n-1}\}$. We may assume that $n \geq 2$ because we may always add letters not appearing in the values of h. For a word

$$x = a_{i_{k-1}} a_{i_{k-2}} \ldots a_{i_0}$$

over Σ, we define

$$\varphi_1(x) = i_0 + i_1 n + \ldots + i_{k-1} n^{k-1} \text{ and } \varphi_2(x) = n^k.$$

Thus, $\varphi_1(x)$ is the integer represented by the word x in the n-ary number system with digits $0, 1, \ldots, n-1$, the letter a_i being viewed as the digit i. The value $\varphi_2(x)$ determines the length k of x. Clearly, x is uniquely determined by the two numbers $\varphi_1(x)$ and $\varphi_2(x)$.

It is now an immediate consequence of the definition of t_1 and t_2 that the assignment of integers

$$v_1 = \varphi_1(h(v)), \quad v_2 = \varphi_2(h(v)), \quad v \in V$$

constitutes a solution for the system P'. Indeed, if $w = w'$ is in S' and, consequently, $h(w) = h(w')$, then the words $h(w)$ and $h(w')$ are of the same length and represent the same n-ary integer. But, for the above assignment, these are the two facts expressed by the equations in P' resulting from $w = w'$.

Since P' and P are equivalent, it follows that the above assignment also satisfies the two equations in P, resulting from the "exceptional" equation $y = y'$. This means that $h(y)$ and $h(y')$ are of the same length and the n-ary numbers represented by them coincide. This is only possible if $h(y) = h(y')$, a contradiction. Thus, we have shown that Goal 2 implies Goal 1.

For instance, let
$$V = \{s, u, v\}, \quad \Sigma = \{a_0, a_1\},$$
$$h(s) = a_1 a_0, \quad h(u) = a_1, \quad h(v) = a_0 a_1 a_1.$$

Assume, further, that

(*) $$suu = uv$$

is an equation in S'. Clearly, $h(suu) = h(uv)$. The polynomial equations corresponding to (*) are

(**) $$s_1 u_2 u_2 + u_1 u_2 + u_1 - u_1 v_2 - v_1 = 0,$$

(**)' $$s_2 u_2 u_2 - u_2 v_2 = 0.$$

The assignment of integers, resulting from the definition of h, is now
$$s_1 = 2, \quad u_1 = 1, \quad v_1 = 3,$$
$$s_2 = 4, \quad u_2 = 2, \quad v_2 = 8.$$

This assignment satisfies (**) and (**)', as it should. Conversely, the fact that h satisfies (*) can be deduced from (**) and (**)'.

We now formulate our last subgoal. We then show that it leads to the preceding subgoal, after which we actually establish the statement of the last subgoal.

Goal 3. Assume that Q is at most a denumerably infinite set of polynomials with integer coefficients and with variables taken from a fixed finite set X. Then there is a finite subset Q' of Q such that every polynomial q in Q can be written in the form

(*) $$q = p_1 q_1 + \ldots + p_m q_m,$$

where each q_i is in Q' and each p_i is a polynomial with integer coefficients and with variables taken from X.

To show that Goal 3 implies Goal 2, consider an arbitrary system P of polynomial equations. Let Q consist of polynomials appearing on the left sides of P, and let Q' be the finite subset of Q whose existence is guaranteed by Goal 3. Finally, let P' be the finite subsystem of P, consisting of equations whose left sides are in Q'. It is now immediate by (*) that every solution of P' is a solution of P.

The proof of Goal 3 is by induction on the cardinality r of the set X of variables. Consider first the basis of induction: $r = 0$. This means that all polynomials are constants. Thus, we have to show the following: every denumerable set Q of integers (not necessarily recursively enumerable) possesses a finite subset Q' such that every integer in Q can be expressed as a linear combination, with integer coefficients, of the integers in Q'.

Consider the smallest positive absolute value $n = |k_0| > 0$ with k_0 in Q. If there are integers in Q not expressible in the form ik_0, where i is an integer, let k_1 have the smallest absolute value among them. It follows that

$$|k_1| > n \text{ and } k_1 \not\equiv 0 \pmod{n}.$$

If there are integers in Q not expressible in the form $ik_0 + jk_1$, let k_2 have the smallest absolute value among them. It follows that k_2 is incongruent to both 0 and k_1 modulo n.

Continuing in the same way, we observe that every new integer must lie in a residue class modulo n different from the residue classes determined by the previous integers. This means that there exists a subset Q' of Q consisting of at most n elements and satisfying our requirements.

Proceeding inductively, we assume that Goal 3 holds true if the set X consists of r variables, and consider a set Q of polynomials with variables from the set $X = \{x_1, \ldots, x_{r+1}\}$. We assume without loss of generality that the set Q is closed under forming linear combinations: whenever $q_1, \ldots, q_k$ are in Q, then so is $p_1 q_1 + \ldots + p_k q_k$, where each p_i is a polynomial with integer coefficients and variables from X. For if this is not the case originally, we replace the original $\bar{Q}$ by such an extended Q. If $q_1, \ldots, q_m$ constitute a basis of Q in the sense of (*), then the finitely many elements of $\bar{Q}$ needed to generate each q_i as a linear combination constitute a basis of $\bar{Q}$.

We denote $x = x_{r+1}$ and write the polynomials q in Q in the form

(**) $$q = p_0 x^s + p_1 x^{s-1} + \ldots + p_{s-1} x + p_s,$$

where each p_i is a polynomial with variables $x_1, \ldots, x_r$ and with integer coefficients. Some p_i may be the zero polynomial. We now apply the inductive hypothesis to the set of polynomials p_0 in (**) when q ranges over Q. This set is generated by finitely many polynomials, say, $p_0^1, \ldots, p_0^e$. Let the corresponding polynomials q (see (**)) be $q^1, \ldots, q^e$. Thus, for instance,

$$q^1 = p_0^1 x^{s_1} + \ldots.$$

Let $\bar{s}$ be the greatest of the exponents s_i, $i = 1, \ldots, e$.

Consider an arbitrary polynomial q as in (**), where the greatest exponent s of x satisfies $s \geq \bar{s}$. We show that q can be written in the form

(***) $$q = p^1 q^1 + \ldots + p^e q^e + q',$$

where the ps and q' are polynomials with integer coefficients and variables from X and, moreover, q' does not contain powers of x with exponents greater than $\bar{s} - 1$. Indeed, we can write p_0 from (*) in the form

$$p_0 = \alpha_1 p_0^1 + \ldots + \alpha_e p_0^e,$$

for some αs. Hence, the greatest exponent of x in the polynomial

$$q - (\alpha_1 x^{s-s_1})q^1 - \ldots - (\alpha_e x^{s-s_e})q^e = q''$$

is $\leq s - 1$. By our assumption, q'' is in Q. If the greatest exponent of x in q'' is still $\geq \bar{s}$, the procedure is repeated with q'' in place of q and carried on until $(***)$ is reached.

We still have to take care of polynomials q', where the greatest exponent of x is $\leq \bar{s} - 1$. For this purpose, we have to extend the basis $\{q^1, \ldots, q^e\}$. The argument follows exactly the same lines as before. By the inductive hypothesis, the set of coefficients of $x^{\bar{s}-1}$ in such polynomials q' is generated by some $r_0^1, \ldots, r_0^f$. Again let $\bar{q}^1, \ldots, \bar{q}^f$ be the corresponding polynomials q. (That is, r_0^1 is the coefficient of $x^{\bar{s}-1}$ in $\bar{q}^1$, etc.) Every q' can be written in the form

$$q' = \bar{p}^1 \bar{q}^1 + \ldots + \bar{p}^f \bar{q}^f + q''',$$

where the greatest exponent of x in q''' is $\leq \bar{s} - 2$. Polynomials q''' are handled in the same way, and the continuation is clear. After at most $\bar{s}$ extensions a finite basis

$$\{q^1, \ldots, q^e, \bar{q}^1, \ldots, \bar{q}^f, \ldots\}$$

for Q results.

We are now ready for a formal statement.

Theorem 2.10. Every language possesses a test set. There is no algorithm for finding a test set for a given recursive language.

Proof. We have already established the first sentence. The second sentence follows because, for instance, it is undecidable whether or not a given recursive language L contains a nonempty word. Let g be the identity morphism and h the morphism mapping all letters to the empty word. If we could effectively find a test set for L, we could decide whether or not g and h are equivalent on L. But we would then be solving the undecidable problem mentioned. □

2.4 Emil's words in retrospect: an overview of PCP

We have elaborated and modified, in various directions, the ideas emanating from the Post Correspondence Problem. Its importance lies in the fact that it is a very suitable reference point for establishing undecidability. Indeed, in a setup involving a pairing of words, reduction to the Post Correspondence Problem is often the most natural way to prove undecidability. Typical examples will be given below.

The purpose of this section is twofold. Firstly we want to illustrate the use of the Post Correspondence Problem in undecidability proofs. This discussion by no means attempts to draw a border line between cases that are suitable and not suitable for a PCP reduction. In fact, such an attempt would be futile because of the translatability between the Post Correspondence Problem and other reference points for establishing undecidability, say the halting problem.

Secondly, we describe the expressive power of the Post Correspondence Problem from a different point of view. We show that in some sense the expressive power exceeds the strength of any formalization. We consider any formal system AX that is consistent and strong enough, meaning that a sufficient amount of arithmetic for carrying out Gödel's argument is included. Otherwise, we may include in AX whatever we want. We then construct an instance PCP_{AX} of the Post Correspondence Problem such that a formal proof within AX is possible neither for the statement "PCP_{AX} has a solution" nor for the statement "PCP_{AX} has no solution". More explicitly, we show that it is possible to represent such statements within the framework of AX — we will return to this question below — and show that the "AX-versions" of the statements are not provable within AX.

At first sight this result is surprising because the idea of a PCP seems simple enough, and we should be able to settle the question of the existence of a solution, even formally, in one way or another. However, the result only shows the expressive power of the Post Correspondence Problem: the possibility of computations and the existence of proofs can be simulated with PCPs. This makes it possible, for example, to construct PCP-instances that essentially assert their own unsolvability.

We now enter the first of the two topics described above: the discussion of some typical PCP-reductions.

A *context-free grammar* can be used to generate correspondences in the following way. Consider a list $(\alpha_1, \ldots, \alpha_n)$ of words. The productions

$$S \to iS\alpha_i, \quad i = 1, \ldots, n,$$

generate arbitrary catenations of the αs to the right of S. The sequence created to the left of S keeps track of the order of the αs. In fact, the words α_i and the numbers i are in reverse order. This is due to the fact that our context-free productions pump both numbers and words from the middle and, consequently, the resulting sequences are reverse to one another when everything is read from left to right.

If one starts with two lists $(\alpha_1, \ldots, \alpha_n)$ and $(\beta_1, \ldots, \beta_n)$, one uses the productions

$$S \to iS\alpha_i, \quad S \to iS\beta_i, \quad i = 1, \ldots, n$$

to generate candidate solutions for a PCP-instance.

Example 2.11. Assume that the two lists are

$$(c^2, d^2, cd^2) \text{ and } (c^2d, dc, d).$$

Consider two context-free grammars G_1 and G_2 with the terminal alphabet $\{1, 2, 3, c, d\}$ and the start symbol S, defined as follows.

$$G_1 : \quad S \to 1Sc^2, \quad S \to 2Sd^2, \quad S \to 3Scd^2, \quad S \to 1c^2, \quad S \to 2d^2, \quad S \to 3cd^2,$$
$$G_2 : \quad S \to 1Sc^2d, \quad S \to 2Sdc, \quad S \to 3Sd, \quad S \to 1c^2d, \quad S \to 2dc, \quad S \to 3d.$$

The productions have been constructed in the way described prior to the example, with additional productions for termination. If the original lists are denoted by αs and βs, we obtain

$$\begin{aligned} L(G_1) &= \{i_1 \ldots i_k \, \alpha_{i_k} \ldots \alpha_{i_1} | k \geq 1 \ , \ 1 \leq i_j \leq 3\}, \\ L(G_2) &= \{i_1 \ldots i_k \, \beta_{i_k} \ldots \beta_{i_1} | k \geq 1 \ , \ 1 \leq i_j \leq 3\}. \end{aligned}$$

Denote by PCP the instance defined by the two lists. Then, obviously,

$$L(G_1) \cap L(G_2) = \emptyset \text{ iff } PCP \text{ has no solution.}$$

In fact, in this case PCP has a solution and

$$L(G_1) \cap L(G_2) = (3121)^i (c^2 d^2 c^3 d^2)^i | i \geq 1\}.$$

□

This idea also works in the general case. Consider an instance PCP determined by the lists

(∗) $\qquad\qquad (\alpha_1, \ldots, \alpha_n) \text{ and } (\beta_1, \ldots, \beta_n).$

We define the grammars G_1 and G_2 as before, with the exception that we encode the indices using the trick given in Example 2.7: in place of i we take ba^i. This is done in order to avoid the increase of the terminal alphabet with the length n. Thus, the grammars have the productions

$$\begin{aligned} G_1 : \quad & S \to ba^i S \alpha_i \quad, \quad S \to ba^i \alpha_i \quad, \quad i = 1, \ldots, n, \\ G_2 : \quad & S \to ba^i S \beta_i \quad, \quad S \to ba^i \beta_i \quad, \quad i = 1, \ldots, n. \end{aligned}$$

(It is assumed that a and b do not belong to the alphabet of the αs and βs.) Now also,

$$L(G_1) \cap L(G_2) = \emptyset \text{ iff } PCP \text{ has no solution.}$$

Hence, if we could decide the emptiness of the intersection of two context-free (or even two linear) languages, we would be deciding the Post Correspondence Problem, which contradicts Theorem 2.1.

We now use the same argument to derive another undecidability result. Roughly, we say that a grammar G is *unambiguous* if no word w in $L(G)$ has two different derivations according to G. Otherwise, G is said to be *ambiguous*. It is not necessary for our purposes to explain the technical details concerning which derivations are regarded as "different". This is clear for linear grammars. Any derivation

$$S \Rightarrow x_1 \Rightarrow \ldots \Rightarrow x_m$$

determines a unique sequence of productions $p_1,\ldots,p_m$ applied in the derivation, unique also as regards the position of application. Two derivations are different iff the associated sequences of productions are different.

We claim that it is undecidable whether or not a given linear grammar G is ambiguous. To prove our claim, we again consider an arbitrary instance PCP determined by the lists (*). Let the productions in a linear grammar G be

$$S \to S_1 \,, \; ; S \to S_2 \,,$$

$$S_1 \to ba^i S_1 \alpha_i \,, \; S_1 \to ba^i \alpha_i \,, \; i = 1, \ldots, n \,,$$

$$S_2 \to ba^i S_2 \beta_i \,, \; S_2 \to ba^i \beta_i \,, \; i = 1, \ldots, n \,.$$

Assume that PCP has a solution $i_1 \ldots i_k$. Then $\alpha_{i_1} \ldots \alpha_{i_k} = \beta_{i_1} \ldots \beta_{i_k}$ and the word

$$ba^{i_k} \ldots ba^{i_1} \alpha_{i_1} \ldots \alpha_{i_k} = ba^{i_k} \ldots ba^{i_1} \beta_{i_1} \ldots \beta_{i_k}$$

has two different derivations according to G, one via S_1 and the other via S_2. Hence, G is ambiguous. On the other hand, this is the only possible ambiguity. If PCP has no solution, then the languages derived via S_1 and S_2 are disjoint and, moreover, no ambiguities occur in S_1- or S_2-derivations. This means that if we can decide whether or not G is ambiguous, we can also decide the Post Correspondence Problem.

Example 2.12. Consider the two lists

$$(c^2, d^2, cd^2) \text{ and } (c^2d, dc, d)$$

from Example 2.11. Consider the grammar G constructed as above. The productions are

$$S \to S_1 \,, \; S \to S_2 \,,$$

$$\begin{array}{llll}
S_1 \to baS_1c^2 & , & S_1 \to ba^2 S_1 d^2 & , & S_1 \to ba^3 S_1 cd^2 \,, \\
S_1 \to bac^2 & , & S_1 \to ba^2 d^2 & , & S_1 \to ba^3 cd^2 \,, \\
S_2 \to baS_2 c^2 d & , & S_2 \to ba^2 S_2 dc & , & S_2 \to ba^3 S_2 d \,, \\
S_2 \to bac^2 d & , & S_2 \to ba^2 dc & , & S_2 \to ba^3 d \,.
\end{array}$$

According to Example 2.10 and the preceding discussion, G should be ambiguous. Indeed, the word $ba^3 baba^2 bac^2 d^2 c^3 d^2$ has two different derivations:

$$S \Rightarrow S_1 \Rightarrow ba^3 S_1 cd^2 \Rightarrow ba^3 baS_1 c^3 d^2 \Rightarrow ba^3 baba^2 S_1 d^2 c^3 d^2$$

$$\Rightarrow ba^3 baba^2 bac^2 d^2 c^3 d^2$$

and

$$S \Rightarrow S_2 \Rightarrow ba^3 S_2 d \Rightarrow ba^3 baS_2 c^2 d^2 \Rightarrow ba^3 baba^2 S_2 dc^3 d^2$$

$$\Rightarrow ba^3 baba^2 bac^2 d^2 c^3 d^2 \,.$$

It is easy to see that no word in $L(G)$ has more than two different derivations, a fact expressed by saying that the *degree of ambiguity* of G equals two. □

We now proceed to another reduction along similar lines. We say that a language L over the alphabet Σ is *thin* iff, with finitely many exceptions, L contains at most one word of each length. More formally,

$$(\exists m_0)(\forall m \geq m_0)(\text{card}(L \cap \Sigma^m) \leq 1).$$

The language L is *p-thin* iff, again with finitely many exceptions, L contains at most p words of each length. Thus, 1-thinness amounts to thinness. Finally, L is *slender* iff

$$(\exists p)(\forall m)(\text{card}(L \cap \Sigma^m) \leq p).$$

Observe here that the lower bound m_0 for the "initial mess" is not needed because it can be taken care of by choosing a larger p.

Assume that we could decide the thinness of the intersection of two linear languages. As before, we consider an arbitrary instance PCP determined by the α- and β-lists (*). We assume that the alphabet of the words in the lists does not contain the letters a, b, c. Define two linear grammars G_1 and G_2 as follows. Both grammars contain the productions

$$S \to S_1,\ S \to S_2,\ S_2 \to cS_2,\ S_1 \to c,\ S_2 \to c.$$

In addition, G_1 (resp. G_2) contains the productions

$$S_1 \to ba^i S_1 \alpha_i\ (\text{resp.}\ S_1 \to ba^i S_1 \beta_i),\ i = 1, \ldots, n.$$

The intersection $L(G_1) \cap L(G_2)$ contains all powers of c and, consequently, it contains a word of each positive length. The intersection contains other words exactly in case the instance PCP has a solution. In this case, because the solution can be iterated, the intersection contains arbitrarily long words other than powers of c. Thus, the intersection $L(G_1) \cap L(G_2)$ is thin iff PCP has no solution. Our assumption has led to the contradiction that we could solve the Post Correspondence Problem.

In the same way we can prove that, for a fixed $p \geq 1$, the p-thinness of the intersection of two linear languages is undecidable. The only difference is that we modify the common part of G_1 and G_2 in such a way that it generates p words of each length m, for instance the powers $c_1^m, \ldots, c_p^m$.

To obtain the same result for slenderness, we modify the productions leading to termination as follows. The basic setup with the instance PCP determined by (*) remains the same. This time we use four additional terminal letters a, b, c, d not appearing in αs and βs when we define two linear grammars G_1 and G_2. Both grammars contain the productions

$$S_1 \to cS_1,\ S_1 \to dS_1,\ S_1 \to c,\ S_1 \to d.$$

In addition, G_1 contains the productions

$$S \to ba^i S_1 \alpha_i\ \text{and}\ S_1 \to ba^i S_1 \alpha_i,\ i = 1, \ldots, n,$$

and G_2 contains the analogous β-productions

$$S \to ba^i S_1 \beta_i \text{ and } S_1 \to ba^i S_1 \beta_i, \quad i = 1, \ldots, n.$$

(Unless stated otherwise, S is the start letter.)

Assume that PCP possesses a solution $i_1 \ldots i_k$. Then the sentential form

$$ba^{i_k} \ldots ba^{i_1} S_1 \alpha_{i_1} \ldots \alpha_{i_k} = u S_1 w$$

can be derived according to both G_1 and G_2. Observe that all words in $\{c, d\}^+$ can be derived from S_1 in both grammars. This implies that

$$u\{c, d\}^+ w \subseteq L(G_1) \cap L(G_2)$$

and, consequently, the intersection cannot be slender.

Conversely, if PCP possesses no solution, then the intersection $L(G_1) \cap L(G_2)$ is empty. This follows because the $\{c, d\}$-parts of eventual words in the intersection must lie in the middle of a solution. Thus, we have shown that $L(G_1) \cap L(G_2)$ is slender iff it is thin iff PCP has no solution. (Observe that this argument can also be used to show the undecidability of thinness and emptiness.)

We now summarize the undecidability results obtained so far in this section.

Theorem 2.11. *It is undecidable whether or not a linear grammar is ambiguous. It is undecidable whether or not the intersection of two linear languages is (i) empty, (ii) thin, (iii) p-thin for a given $p \geq 1$, (iv) slender.*

In the preceding constructions, we attempted to generate solutions for PCP. We now use the technique of generating nonsolutions. It will be convenient for our purposes to use the encoding presented in Example 2.7 and assume that, in the instance PCP determined by the lists (*), the words α_i and β_i are over the alphabet $\{a, b\}$. In fact, we also use the same alphabet to encode the indices i. No confusion will arise because the two encodings are separated by a marker c: we consider words of the form

(**) $\qquad ba^{i_k} \ldots ba^{i_1} c \alpha_{i_1} \ldots \alpha_{i_k}, \quad k \geq 1.$

For a list $(\alpha_1, \ldots, \alpha_n)$, we now construct a linear grammar G_α such that $L(G_\alpha)$ equals the subset of

$$L(a, b, c, n) = \{ba, ba^2, \ldots, ba^n\}^* c \{a, b\}^*,$$

consisting of those words that are *not* of the form (**). The nonterminals of G_2 are A', A, A_1, A_2, A_3, with A' being the start letter. (Eventually, G_α will be included in a bigger grammar.)

Thus, we want to generate the language L_α consisting of those words of $L(a, b, c, n)$ that are not of the form (**). We first list the productions of G_α:

$$A' \to A, \ A' \to c \ ,$$
$$A \to ba^i A \alpha_i \ , \quad i = 1, \ldots, n \ ,$$
$$A \to ba^i A_1 x \ , \quad i = 1, \ldots, n \ ,$$

where x ranges over those words in $\{a,b\}^*$ that are shorter than α_i,

$$A \to ba^i A_2 x \ , \quad i = 1, \ldots, n \ ,$$

where x ranges over those words in $\{a,b\}^*$ satisfying $x \neq \alpha_i$ and $|x| = |\alpha_i|$,

$$A_1 \to ba^i A_1 \ , \ A_2 \to ba^i A_2 \ , \ i = 1, \ldots, n \ ,$$

$$A_1 \to c \ , \ A_2 \to A_3 \ ,$$

$$A \to A_3 a \ , \ A \to A_3 b \ , \ A_3 \to A_3 a \ \ A_3 \to A_3 b \ , \ A_3 \to c \ .$$

We show that $L(G_\alpha) = L_\alpha$. Consider words in $L(G_\alpha)$. Termination is possible only from A_1 and A_3. (We disregard the productions $A' \to A$ and $A' \to c$ needed only for the word c.) If it happens from A_1, the sequence of nonterminals used is $A, \ldots, A, A_1, \ldots, A_1$, and in the resulting word the subword to the right of c is too short for (**). If termination happens from A_3, the sequence of nonterminals used in the derivation is $A, \ldots, A, A_2, \ldots, A_2, A_3, \ldots, A_3$, where an irreparable inconsistency with (**) was created in the transition from A to A_2. Since, clearly, all words generated by G_α are in $L(a,b,c,n)$, we conclude that $L(G_\alpha)$ is included in L_α.

To establish the reverse inclusion, we consider an arbitrary word w in L_α. If w is in $c\{a,b\}^*$, w can be generated via the nonterminals A and A_3. Otherwise, w is of the form

$$w = ba^{i_k} \ldots ba^{i_1} c w' \ , \ k \geq 1 \ ,$$

where $w' \neq \alpha_{i_1} \ldots \alpha_{i_k}$, $w' \in \{a,b\}^*$. Let l be the smallest integer, if any, in the interval $1 \leq l \leq k$ such that $w' = w_1 \alpha_{i_l} \ldots \alpha_{i_k}$. (Thus, we take the longest suffix of w' consistent with (**).)

Using productions $A \to ba^i A \alpha_i$, we reach the word (sentential form)

$$ba^{i_k} \ldots ba^{i_l} A \alpha_{i_l} \ldots \alpha_{i_k} \ .$$

(This step is unnecessary if no l exists as required.) If $l = 1$, w_1 must be nonempty because, otherwise, w would be of the form (**). We may now generate w_1 and also w using first one of the productions $A \to A_3 a$ or $A \to A_3 b$, and then productions for A_3. If $l > 1$, the choice of l guarantees that $\alpha_{i_{l-1}}$ is not a suffix of w_1. Two subcases arise. If $|w_1| < |\alpha_{i_{l-1}}|$, then we use first the production

$$A \to ba^{i_{l-1}} A_1 w_1 \ ,$$

after which we reach w by applying productions $A_1 \to ba^i A_1$ and $A_1 \to c$. If $|w_1| \geq |\alpha_{i_{l-1}}|$, we write $w_1 = w_2 x$, where $|x| = |\alpha_{i_{l-1}}|$. The production $A \to ba^{i_{l-1}} A_2 x$ is applied to our sentential form, after which we use productions $A_2 \to ba^i A_2$ (needed in case $l > 2$), $A_2 \to A_3$ and, finally, reach w by productions for A_3.

We have shown that in all cases w is in $L(G_\alpha)$ and, consequently, $L_\alpha = L(G_\alpha)$.

The language L_β is defined analogously with respect to the list $(\beta_1, \ldots, \beta_n)$: L_β is the subset of $L(a, b, c, n)$, consisting of those words that are not of the form

$$ba^{i_k} \ldots ba^{i_1} c \beta_{i_1} \ldots \beta_{i_k}, \quad k \geq 1.$$

The construction of a linear grammar G_β for L_β is of course exactly analogous. The nonterminals of G_β are denoted B(start), B_1, B_2, B_3. (The additional B' is not needed here.)

It is obvious that

$$L(G_\alpha) \cup L(G_\beta) = L(a, b, c, n) \text{ iff } PCP \text{ has no solution}.$$

Example 2.13. We return to Example 2.12, and write the lists according to our convention concerning the alphabet:

$$(a^2, b^2, ab^2) \text{ and } (a^2 b, ba, b).$$

The grammar G_α will consist of the productions:

$A' \to A$, $A' \to c$,
$A \to baAa^2$, $A \to ba^2 Ab^2$, $A \to ba^3 Aab^2$,
$A \to baA_1 a$, $A \to baA_1 b$, $A \to baA_1$,
$A \to ba^2 A_1 a$, $A \to ba^2 A_1 b$, $A \to ba^2 A_1$,
$A \to ba^3 A_1 x$, with $x = a^2, ab, ba, b^2, a, b, \lambda$,
$A \to baA_2 x$, with $x = ab, ba, b^2$,
$A \to ba^2 A_2 x$, with $x = a^2, ab, ba$,
$A \to ba^3 A_2 x$, with $x = a^3, a^2 b, aba, ba^2, bab, b^2 a, b^3$,
$A_1 \to baA_1$, $A_1 \to ba^2 A_1$, $A_1 \to ba^3 A_1$,
$A_2 \to baA_2$, $A_2 \to ba^2 A_2$, $A_2 \to ba^3 A_2$,
$A_1 \to c$, $A_2 \to A_3$,
$A \to A_3 a$, $A \to A_3 b$, $A_3 \to A_3 a$, $A_3 \to A_3 b$, $A_3 \to c$.

The definition of G_β is analogous and omitted. The word $ba^3 baca^2 b^2$ is properly constructed from the β-list and, hence, cannot be derived according to G_β. It can be derived in G_α:

$$A \Rightarrow ba^3 Aab^2 \Rightarrow ba^3 baA_1 aab^2 \Rightarrow ba^3 baca^2 b^2.$$

The instance PCP that we are considering possesses a solution 1213 and, hence, the resulting word

$$ba^3 baba^2 baca^2 b^2 a^3 b^2$$

is derivable according to neither G_α nor G_β. This means that $L(G_\alpha) \cup L(G_\beta)$ is properly contained in $L(a, b, c, n)$. □

The grammars G_α and G_β generate words that are "wrong" for the α- and β-lists. Their use in PCP-reductions is illustrated in the following theorem. The terms *co-thin* and *co-slender* refer to complements: a language is co-thin (resp. co-slender) iff its complement is thin (resp. slender).

Theorem 2.12. Each of the following problems is undecidable:

(i) Do two given linear grammars generate the same sentential forms?
(ii) Do two given linear grammars generate the same language?
(iii) Are a given linear and a given regular language equal?
(iv) Do two given OL systems generate the same language?
(v) Is a linear language co-thin?
(vi) Is a linear language co-slender?

Proof. Throughout the proof, we have in mind the instance PCP determined by the lists $(\alpha_1, \ldots, \alpha_n)$ and $(\beta_1, \ldots, \beta_n)$, as well as the grammars G_α and G_β defined as above.

(i) We construct two linear grammars G_1 and G_2. Both have the start letter S and an additional nonterminal S_1.

The grammar G_1 contains all productions of G_α and G_β and, in addition, the productions

$$\begin{aligned} S &\to A', S \to B, S \to S_1, \\ S_1 &\to ba^i S_1, \ i = 1, \ldots, n, \\ S_1 &\to S_1 a, \ S_1 \to S_1 b. \end{aligned}$$

The grammar G_2 is obtained from G_1 by replacing the terminal productions

$$A_1 \to c, \ A_3 \to c, \ B_1 \to c, \ B_3 \to c$$

with the single production $S_1 \to c$.

It is immediate by the definition of G_1 and G_2 that the only possible difference in the sets of sentential forms generated by them lies in the words over the terminal alphabet $\{a, b, c\}$. In other words, G_1 and G_2 generate the same sentential forms iff $L(G_1) = L(G_2)$. But $L(G_1) = L(G_\alpha) \cup L(G_\beta)$ because the productions involving S_1 do not contribute to the language of G_1. It is also clear that $L(G_2) = L(a, b, c, n)$ because, according to G_2, all terminal words are generated via S_1. To decide whether or not G_1 and G_2 generate the same sentential forms amounts to deciding whether or not

$$L(G_\alpha) \cup L(G_\beta) = L(a, b, c, n)$$

which, in turn, amounts to deciding whether PCP has a solution.

(ii)–(iv) We have actually already settled (ii), and (iii) follows because $L(G_2)$ is regular. To prove (iv), we transform G_1 and G_2 into OL systems H_1 and H_2 by adding the productions $a \to a$, $b \to b$, $c \to c$ and disregarding the difference between terminals and nonterminals. Then the OL languages $L(H_1)$ and $L(H_2)$

coincide with the sets of sentential forms of G_1 and G_2 respectively. This shows that an algorithm for deciding whether or not $L(H_1) = L(H_2)$ yields an algorithm for deciding whether or not PCP has a solution.

(v) Consider the regular language

$$R = \sim L(a,b,c,n) \cap \sim c^* = \sim \{ba,\ldots,ba^n\}^* c\{a,b\}^* \cap \sim c^*.$$

Let G_3 be a linear grammar generating the language

$$L(G_1) \cup R = L_\alpha \cup L_\beta \cup R.$$

Consequently,

$$\sim L(G_3) = \sim (L_\alpha \cup L_\beta) \cap \sim R = \sim (L_\alpha \cup L_\beta) \cap (L(a,b,c,n) \cup c^*).$$

This means that all words c^i, $i \neq 1$, are in $\sim L(G_3)$ and that $\sim L(G_3)$ contains other words exactly in case $L_\alpha \cup L_\beta$ is strictly included in $L(a,b,c,n)$. In this case there are infinitely many other words in $\sim L(G_3)$. This implies that $L(G_3)$ is co-thin exactly in case PCP possesses no solution.

(vi) Here we use a technique similar to the one used in proving the undecidability of the intersection of two linear languages being slender. Roughly, the marker c is replaced by the set of all nonempty words over $\{c,d\}$.

Denote

$$L' = L'(a,b,c,d,n) = \{ba,\ldots,ba^n\}^* \{c,d\}^+ \{a,b\}^*.$$

Let L'_α be the subset of L', consisting of those words that are *not* of the form

$$ba^{i_k}\ldots ba^{i_1} w \alpha_{i_1}\ldots \alpha_{i_k},\ k \geq 1,\ w \in \{c,d\}^+.$$

The definition of L'_β is analogous. It follows that $L'_\alpha \cup L'_\beta = L'$ iff PCP has no solution. A linear grammar G'_1 generating $L'_\alpha \cup L'_\beta$ is obtained from G_1 by replacing everywhere the terminal c with the nonterminal C and adding the productions

$$C \to cC,\ C \to dC,\ C \to c,\ C \to d.$$

(Because we have used G_1 for many purposes, G'_1 so defined contains some productions not needed for (vi). However, they are harmless.) We choose, finally, $R' = \sim L'$ and a linear grammar G'_3 such that

$$L(G'_3) = L(G'_1) \cup R' = L'_\alpha \cup L'_\beta \cup R'.$$

Here the situation is quite conspicuous: the complement of $L(G'_3)$ equals the language

$$L' - (L'_\alpha \cup L'_\beta)$$

that is nonempty and, at the same time, non-slender exactly in case PCP possesses a solution. □

So far in this section we have illustrated the use of the Post Correspondence Problem in undecidability proofs. The topics chosen range from classical to very recent. To complete our overview of Emil's words in retrospect, we now enter our second topic. We describe the strength of PCPs by showing that statements of the form "the instance PCP has a solution" escape every formal system.

More specifically, we consider any consistent formal system AX that includes addition and multiplication of integers. Given AX, we construct an instance PCP_{AX} of the Post Correspondence Problem such that neither of the statements "PCP_{AX} has a solution" or "PCP_{AX} has no solution" is provable within AX. Thus, statements such as this somehow go beyond the formal system. This can be explained as follows. Statements about an instance PCP_{AX} having or not having a solution can be "expressed" or "represented" within the formal system but cannot be settled within the formal system.

The arguments given below are only given in outline. Detailed constructions, say within formalized arithmetic, would be tedious and very lengthy. We assume from now on that AX has been fixed. Provability within AX will be denoted by the symbol $\vdash$.

We illustrate first how statements about an instance PCP having or not having a solution can be expressed within the formal system AX. In fact, the transition from word equations to numerical equations has been discussed already in connection with Goal 2 in the proof of Theorem 2.10.

Example 2.14. Consider again the instance PCP determined by the lists

$$(a^2, b^2, ab^2) \text{ and } (a^2b, ba, b),$$

or, equivalently, by the morphisms $g, h : \{A, B, C\}^* \to \{a, b\}^*$

$$g(A) = a^2 \; , \; g(B) = b^2 \; , \; g(C) = ab^2 \; ,$$
$$h(A) = a^2b \; , \; h(B) = ba \; , \; h(C) = b \; .$$

The word $ABAC$ being a solution means that the equation

$$ABAC = \bar{A}\bar{B}\bar{A}\bar{C}$$

is satisfied when the letters (resp. barred letters) are replaced by the corresponding g-value (resp. h-value). (This holds true because the word $a^2b^2a^3b^2$ results from both sides.) Using the two functions t_1 and t_2 from the proof of Theorem 2.10, this word equation is replaced by two numerical equations

$$A_1B_2A_2C_2 + B_1A_2C_2 + A_1C_2 + C_1 = \bar{A}_1\bar{B}_2\bar{A}_2\bar{C}_2 + \bar{B}_1\bar{A}_2\bar{C}_2 + \bar{A}_1\bar{C}_2 + \bar{C}_1 \; ,$$

$$A_2B_2A_2C_2 = \bar{A}_2\bar{B}_2\bar{A}_2\bar{C}_2 \; .$$

The values assigned to the letters indexed by 1 are the g- and h-values, viewed as binary integers ($a = 0, b = 1$):

$$A_1 = 0 \, , \, B_1 = C_1 = 3 \, , \, \bar{A}_1 = \bar{C}_1 = 1 \, , \, \bar{B}_1 = 2 \, .$$

The values assigned to the letters indexed by 2 are powers of 2, where the exponent indicates the length of the g- or h-value:

$$A_2 = B_2 = 4,\ C_2 = 8,\ \bar{A}_2 = 8,\ \bar{B}_2 = 4,\ \bar{C}_2 = 2.$$

This assignment satisfies the numerical equations, as it should!

Thus, the statement that a word w is a solution for PCP can be expressed as two polynomial equations (reducible to one equation). The latter result from w by the definition of g and h.

The existence of a solution w can be expressed as the existence of a number x from which w results. For instance, the encoding might be as follows. Assume that $\Sigma = \{a_1, \ldots, a_n\}$ (the domain alphabet of g and h) and that $w = a_{i_1} \ldots a_{i_k}$. Then

$$x = p_1^{i_1} \ldots p_k^{i_k},$$

where the ps are the first k primes. Thus, the exponent gives the letter and the prime its position in w. By this encoding, our above solution $ABAC$ is read from the number

$$x = 2^1 \cdot 3^2 \cdot 5^1 \cdot 7^3 = 30\,870.$$

$\square$

Given an instance PCP determined by the morphisms g and h, we define the predicate

$$SOLUTION_{g,h}(x)$$

as an abbreviation for the conjunction of the following conditions:

(i) x is of the proper form, that is, a product of the first k primes, raised to some positive powers;
(ii) if $w = a_{i_1} \ldots a_{i_k}$ is the word read from the exponents, then the numerical equations corresponding to the word equation $w = \bar{w}$ (as explained in Example 2.13) are satisfied.

Thus, the existence of a solution for PCP is expressed as

$$(\exists x) SOLUTION_{g,h}(x),$$

and the nonexistence of a solution is expressed as its negation

$$\neg(\exists x) SOLUTION_{g,h}(x).$$

Our purpose is to construct an instance PCP_{AX} such that neither of these statements is provable within AX, when the solution predicate is defined using the morphisms from PCP_{AX}.

The computational model that we use in the following is the Turing machine. Normal systems or any other model capable of producing all recursively enumerable sets would do as well. We use Turing machines because they are most intuitive.

For a Turing machine TM and input w, the predicate $HALT(TM,w)$ is true iff TM halts with w. The predicate can be expressed as a numerical predicate representable in AX. We also need the predicate $THEOREM(\alpha)$, also representable in AX, meaning that a well-formed formula α is provable within AX.

We are now in a position to state three formalization results.

Formalization result 1. For a Turing machine TM and input w, we may construct an instance $PCP = (g, h)$ of the Post Correspondence Problem such that

$$\vdash (HALT(TM,w) \longleftrightarrow (\exists x)SOLUTION_{g,h}(x)).$$

Proof idea. We consider a reduction, analogous to the one presented in Theorem 2.1, of the halting problem to the Post Correspondence Problem. The steps in this reduction can be represented within AX.

Formalization result 2. For a well-formed formula α (in the language of AX), we may construct a Turing machine TM and an input w such that

$$\vdash (THEOREM(\alpha) \longleftrightarrow HALT(TM,w)).$$

Proof idea. TM checks the proofs of AX and accepts the last lines of valid proofs. (It is convenient to change the input format from α to w.) Again, the various steps can be represented within AX. □

Our third result is an immediate corollary of the two preceding results.

Formalization result 3. For a well-formed formula α, we may construct an instance $PCP = (g, h)$ of the Post Correspondence Problem such that

(∗) $\qquad \vdash (THEOREM(\alpha) \longleftrightarrow (\exists x)SOLUTION_{g,h}(x)),$

(∗∗) $\qquad \vdash (\neg THEOREM(\alpha) \longleftrightarrow \neg(\exists x)SOLUTION_{g,h}(x)).$

We now choose α to be a Gödel sentence for AX. (See Theorem 1.18.) Consequently, since we have assumed AX to be consistent,

$$\text{not } \vdash \alpha \text{ and not } \vdash \neg\alpha.$$

Let $PCP = (g, h)$ be an instance of the Post Correspondence Problem, constructed by formalization result 3. We are now ready for our main result.

Theorem 2.13. The instance (g, h) constructed above qualifies for PCP_{AX}, that is,

(∗)′ $\qquad \text{not } \vdash (\exists x)SOLUTION_{g,h}(x)$

and

$(**)'$ $\qquad$ not $\vdash \neg(\exists x)SOLUTION_{g,h}(x)$.

Proof. Assume first that $(*)'$ does not hold:

$$\vdash (\exists x)SOLUTION_{g,h}(x) .$$

We obtain from (*) (using propositional logic)

$$\vdash THEOREM(\alpha) ,$$

which is not consistent with our assumption:

$$\text{not } \vdash \alpha .$$

Assume, secondly, that $(**)'$ does not hold:

$$\vdash \neg(\exists x)SOLUTION_{g,h}(x) .$$

We obtain from (**)

$(***)$ $\qquad$ $\vdash \neg THEOREM(\alpha)$.

The consistency of AX can be expressed by saying that the well-formed formula

$$0 = 1$$

is not provable. We abbreviate $\neg THEOREM(0=1)$ by CON. We also obtain

$(***)'$ $\qquad$ $\vdash (THEOREM(0=1) \to THEOREM(\alpha))$.

Indeed, anything can be proved by propositional logic from a contradiction. This proof-theoretic argument can be represented in AX.

By propositional logic,

$$\vdash [(THEOREM(0=1) \to THEOREM(\alpha)) \to (\neg THEOREM(\alpha) \to CON)] .$$

We now obtain from $(***)'$ by *modus ponens*

$$\vdash (\neg THEOREM(\alpha) \to CON)$$

and, hence, from (***) by *modus ponens*

$$\vdash CON .$$

This implies, by Theorem 1.19, that AX is inconsistent.

□

Chapter 3
Diophantine Problems

3.1 Reduction to machine models

In this chapter we will discuss another reference point for establishing the undecidability of given problems — reference point being understood here in the same sense as at the beginning of the preceding chapter. The reference point discussed in this chapter belongs to an area that is very central to mathematics, namely, arithmetic of integers. It is a very good illustration of how "everyday considerations" in mathematics may lead to undecidability. Of course, such a state of affairs is basically due to the expressive power of the problem we will be considering in this chapter. This problem, *Hilbert's Tenth Problem*, can justly be referred to as the most famous specific decision problem. The problem, proposed by Hilbert in a list of problems at the International Congress of Mathematicians in 1900, is to propose an algorithm that will determine whether or not a polynomial equation with integer coefficients has a solution in integers.

When commenting at the same Congress on the proof of the impossibility of solving certain mathematical problems, such as squaring the circle or solving equations of the fifth degree in radicals, Hilbert came to the conclusion that although the problems had no solutions in the sense originally intended, a precise and completely satisfactory way out had still been found. Such a "precise and completely satisfactory" solution to Hilbert's Tenth Problem was provided by Matijasevitch in 1970: undecidable. Nothing was known about undecidability in 1900 and, in view of Hilbert's later attempts to find general decision methods in logic, it remains questionable whether in 1900 he really had in mind the possibility of a negative solution for the Tenth Problem.

Thus, we consider equations of the form

$$(*) \qquad P(x_1, \ldots, x_n) = 0 \,,$$

where P is a polynomial with integer coefficients in the variables $x_1, \ldots, x_n$. From the point of view of decidability it is irrelevant whether we are looking for integer or nonnegative integer solutions for (∗). This can be seen from the following argument.

Every nonnegative integer is a sum of four squares and, consequently, the equation (∗) has a solution in nonnegative integers iff the equation

$$P(p_1^2 + q_1^2 + r_1^2 + s_1^2, \ldots, p_n^2 + q_n^2 + r_n^2 + s_n^2) = 0$$

has an integral solution. On the other hand, (∗) has a solution in integers iff the equation

$$P(p_1 - q_1, \ldots, p_n - q_n) = 0$$

has a solution in nonnegative integers. Thus, an algorithm for one of the problems (finding a solution for an arbitrary equation in integers or in nonnegative integers) can immediately be converted into an algorithm for the other problem.

Hilbert's Tenth Problem will be discussed in Sections 3.1 and 3.2. Some reduction examples and typical techniques will be presented in Section 3.3. Things become much simpler if exponential terms are allowed in (∗), that is, if exponential Diophantine representations are considered. This will be our starting point.

We now begin the formal details. It will be convenient for us to look for solutions in nonnegative integers, rather than in arbitrary integers.

Definition. A set S of ordered n-tuples $(a_1, \ldots, a_n)$, $n \geq 1$, of nonnegative integers is termed *Diophantine* iff there are polynomials $P(a_1, \ldots, a_n, x_1, \ldots, x_m)$ and $Q(a_1, \ldots, a_n, x_1, \ldots, x_m)$ with nonnegative integer coefficients such that, for all nonnegative integers $a_1, \ldots, a_n$,

$$(a_1, \ldots, a_n) \in S \text{ iff } (\exists x_1 \geq 0) \ldots (\exists x_m \geq 0)$$

$$[P(a_1, \ldots, a_n, x_1, \ldots, x_m) = Q(a_1, \ldots, a_n, x_1, \ldots, x_m)].$$

Similarly, S is termed *exponential Diophantine* iff exponentiation is allowed in the formation of P and Q, that is, P and Q are functions built up from $a_1, \ldots, a_n$, $x_1, \ldots, x_m$ and nonnegative integer constants by the operations of addition $A + B$, multiplication AB and exponentiation A^B.

□

Some clarifying remarks are in order. For $n > 1$ we usually speak of *Diophantine* and *exponential Diophantine relations*. In the Diophantine case an equivalent definition results if only one polynomial P and the equation

$$P(a_1, \ldots, a_n, x_1, \ldots, x_m) = 0$$

are considered, provided the coefficients of P may be negative integers. The terms *unary* and *singlefold* are sometimes used in connection with exponential Diophantine representations. Here *singlefold* means that, for any n-tuple $(a_1, \ldots, a_n)$, there is at most one m-tuple $(x_1, \ldots, x_m)$ satisfying the equation. *Unary* means that only one-place exponentials, such as 2^x, are used rather than two-place exponentials y^x. One of the first observations given below will be that it is no loss of generality to restrict our attention to unary representations.

Reduction to machine models

A *Diophantine equation* is a polynomial equation with integer coefficients, in which we are interested only in integer solutions or, as will be the case below, in nonnegative integer solutions. The theory of such equations is very old (it goes back to the Greeks) in classical mathematics. The classical approach is to begin with an equation and ask questions about the set of solutions. However, as Martin Davis has observed, in decidability theory and generally in logic, the classical procedure is turned around. We begin with an interesting set or relation, such as the set of primes or the relation "x is the greatest common divisor of y and z", and look for a Diophantine or an exponential Diophantine equation representing it, in the sense of the definition given above.

It follows immediately by the definition that every Diophantine set or relation is exponential Diophantine. Moreover, exponential Diophantine sets and relations are recursively enumerable. It turns out that the converses of these statements also hold true. The purpose of this section is to establish the converse of the latter statement, often referred to as the *Davis–Putnam–Robinson Theorem*.

Theorem 3.1. All recursively enumerable sets and relations are exponential Diophantine.

The proof of Theorem 3.1 is based on a sequence of constructions, each one showing that a set or relation is exponential Diophantine, abbreviated ED. Many of our constructions actually show more: the set considered is Diophantine. In view of the results given in Section 3.2, we will not emphasize this irrelevant distinction. All variables range over the set of nonnegative integers — this *convention* should be kept in mind.

There are various straightforward techniques for showing sets to be ED. The language of ED sets permits the use of existential quantifiers, this being clear from the definition. For instance, if a relation $R(a_1, \ldots, a_n)$ is ED, then so is the $(n-1)$-place relation $(\exists a_n) R(a_1, \ldots, a_{n-1}, a_n)$. Similarly, conjunction and disjunction may be used in the language of ED sets. This follows because

$$(P_1 = 0 \text{ and } P_2 = 0) \quad \text{iff} \quad P_1^2 + P_2^2 = 0,$$
$$(P_1 = 0 \text{ or } P_2 = 0) \quad \text{iff} \quad P_1 P_2 = 0.$$

On the other hand, the language of ED sets permits neither the use of universal quantifiers nor the use of logical negation.

The key idea in the proof given below is to show that a certain "dominance" relation is ED, and to use the resulting bounded universal quantification to represent arbitrarily long computations of a register machine.

We first show that certain simple number-theoretic relations are ED. We use the customary notations $a|b$ for a divides b, and $\text{rem}(a, b)$ for the remainder after a is divided by b.

Lemma 3.2. The binary relations $a < b$, $a \leq b$, $a \neq b$, $a|b$, as well as the ternary relations $c = \text{rem}(a, b)$ and $a \equiv b \pmod{c}$, are all ED.

Proof. ED-representations can be given as follows:

$$
\begin{aligned}
a < b & \quad \text{iff} \quad (\exists x)(a + x + 1 = b)\,, \\
a \leq b & \quad \text{iff} \quad (\exists x)(a + x = b)\,, \\
a \neq b & \quad \text{iff} \quad (a < b \text{ or } b < a)\,, \\
a | b & \quad \text{iff} \quad (\exists x)(b = ax)\,, \\
a \equiv b \,(\mathrm{mod}\, c) & \quad \text{iff} \quad (\exists x)(a = b + cx \text{ or } b = a + cx)\,, \\
c = \mathrm{rem}(a, b) & \quad \text{iff} \quad (a \equiv c \,(\mathrm{mod}\, b) \text{ and } c < b)\,.
\end{aligned}
$$

□

In fact, these representations show that the relations are singlefold ED. Singlefoldedness will be important in Chapter 5. Observe also that two-place exponentials can always be replaced by one-place exponentials using the formula

$$\mathrm{rem}(2^{ab^2}, 2^{ab} - a) = a^b\,,$$

valid for $1 < b$. Indeed, because $2^{ab} \equiv a \,(\mathrm{mod}\, 2^{ab} - a)$, we infer

$$2^{ab^2} = (2^{ab})^b \equiv a^b \,(\mathrm{mod}\, 2^{ab} - a)\,,$$

from which the claim follows because $a^b < 2^{ab} - a$.

The next lemma is a very important tool.

Lemma 3.3. *The ternary (binomial coefficient) relation $m = \binom{n}{k}$ is ED.*

Proof. Consider first the definition of binomial coefficients in terms of the equation

$$(*) \qquad (u+1)^n = \sum_{i=0}^{n} \binom{n}{i} u^i\,.$$

This is valid for any u. The idea now is to choose u large enough so that $(*)$ can be viewed as the representation of the number $(u+1)^n$ in the number system with base u. Thus, the binomial coefficients $\binom{n}{i}$ are digits in this representation. Since $2^n \geq \binom{n}{i}$, for all i, any u satisfying $u > 2^n$ is large enough. We can make u unique by choosing $u = 2^n + 1$. Clearly, the digits are unique in the base u representation of any number. Consequently, by $(*)$, $m = \binom{n}{k}$ holds iff m is the unique coefficient of u^k in the base u representation of $(u+1)^n$. In this representation we combine the higher powers of u to the form wu^{k+1}, the lower powers being combined simply as a term v with $v < u^k$. Here w and v are unique.

To summarize, $m = \binom{n}{k}$ holds true iff there are unique nonnegative integers u, w and v such that each of the following conditions is satisfied:

$$u = 2^n + 1,\ (u+1)^n = wu^{k+1} + mu^k + v,\ v < u^k,\ m < u\,.$$

Reduction to machine models

By Lemma 3.2 and because we may form conjunctions, we conclude that $m = \binom{n}{k}$ is ED.

At this stage it is still possible to write down explicitly a readable ED representation for the relation we are considering. So let us do it. In the case of our more general results, readability is lost or at least almost lost if we try to achieve an explicit ED representation.

We only have to replace the connective *and* by the squaring technique, and to eliminate the inequalities by introducing new variables. Consequently, the relation $m = \binom{n}{k}$ holds iff

$$(\exists u)(\exists w)(\exists v)(\exists x)(\exists y)[(2^n + 1 - u)^2 + (u^k - v - x)^2$$
$$+ (u - m - y)^2 + ((u+1)^n - wu^{k+1} - mu^k - v)^2 = 0].$$

We have observed that u, w, v, x, y are unique when they exist and, hence, this representation is singlefold. This observation is irrelevant for our present purposes but will become important in Chapter 5.

If we want to have an ED representation exactly in accordance with the definition given above, we have to eliminate the negative coefficients. This is easily accomplished by writing the equation in brackets in the form

$$2^{2n} + 1 + 2u^2 + 2^{n+1} + u^{2k} + v^2 + x^2 + 2vx + m^2 + y^2 + 2my + (u+1)^{2n}$$
$$+ w^2 u^{2k+2} + m^2 u^{2k} + v^2 + 2vwu^{k+1} + 2vmu^k + 2mwu^{2k+1}$$
$$= 2u + 2^{n+1}u + 2vu^k + 2xu^k + 2mu + 2yu + 2w(u+1)^n u^{k+1}$$
$$+ 2m(u+1)^n u^k + 2v(u+1)^n.$$

$\square$

Example 3.1. Let us verify the ED representation given in the proof above for the simple case $\binom{4}{2} = 6$. Thus, $n = 4$, $k = 2$ and $m = 6$. Now $u = 2^4 + 1 = 17$, and we are interested in the base u representation of the number

$$(u+1)^n = 18^4 = 104\,976.$$

Indeed,
$$104\,976 = 1 \cdot 17^4 + 4 \cdot 17^3 + 6 \cdot 17^2 + 4 \cdot 17 + 1.$$

This shows that
$$w = 1 \cdot 17 + 4 = 21 \quad \text{and} \quad v = 4 \cdot 17 + 1 = 69,$$

and consequently,
$$x = 17^2 - 69 = 220 \quad \text{and} \quad y = 17 - 6 = 11.$$

It can be immediately verified that the equations presented in the proof of Lemma 3.3 are satisfied for these values of m, n, k, u, w, v, x and y.

□

We will now define the most important tool, the *dominance relation*, used in the ED representation of register machine computations. Consider two nonnegative integers a and b, written in the binary notation

$$a = \sum_{i=0}^{n} a_i 2^i \ , \quad b = \sum_{i=0}^{n} b_i 2^i \quad (0 \leq a_i, b_i \leq 1) \, .$$

(If necessary, we introduce initial 0s to obtain the same n for both a and b.) By definition, b *dominates* a, in symbols $a \leq_D b$ iff $a_i \leq b_i$, for all $i = 0, \ldots, n$.

Clearly, the dominance relation is a partial order, and $a \leq_D b$ implies $a \leq b$. The converse implication is not valid: $3 = 011$ is not dominated by $4 = 100$.

Lemma 3.4. $a \leq_D b$ iff $\binom{b}{a} \equiv 1 \pmod{2}$. Consequently, the dominance relation is ED.

Proof. The second sentence follows from the first by Lemmas 3.3 and 3.2. Consider the first sentence. It follows by a theorem of Lucas from 1878 — the result was also known to Kummer in the 1850s. Here we need only a special case, in which the proof is fairly short.

As before, let a_i and b_i, $0 \leq i \leq n$, be the bits in the binary representations of a and b. We assume that $a < b$, the first sentence being obvious for $a \geq b$. It suffices to establish the following.

Claim. $\binom{b}{a}$ is even iff $b_i < a_i$, for some i.

Let us denote by $ONE(c)$ the number of 1s in the binary representation of an integer c. The notation $EXP(c)$ is used for the exponent of the highest power of 2 dividing c. Thus, c is odd exactly in case $EXP(c) = 0$. We also obtain, for any $c \geq 1$,

(∗) $$EXP(c!) = c - ONE(c) \, .$$

Indeed, (∗) holds for $c = 1$, both sides being equal to 0. We proceed inductively. Assuming that (∗) holds, we consider the value $c + 1$. If c is even,

$$EXP((c+1)!) = EXP(c!) = c - ONE(c) = (c+1) - ONE(c+1) \, ,$$

because $ONE(c+1) = ONE(c) + 1$. If c is odd, we obtain

$$EXP(c+1) = ONE(c) - ONE(c+1) + 1 \, .$$

($EXP(c+1)$ equals the number of final 0s in the binary expansion of $c+1$, this number depending on the number of final 1s in the binary expansion of c.) Hence, by $(*)$,
$$EXP((c+1)!) = EXP(c!) + EXP(c+1)$$
$$= c - ONE(c) + ONE(c) - ONE(c+1) + 1 = c + 1 - ONE(c+1)$$
This completes the induction, and $(*)$ follows.

Since $\binom{b}{a} = b!/a!(b-a)!$, we now obtain

$$EXP\left(\binom{b}{a}\right) = b - ONE(b) - a + ONE(a) - (b-a) + ONE(b-a)$$
$$= ONE(a) + = ONE(b-a) - ONE(b).$$

Thus, our Claim can be stated in the form

$(**)$ $\qquad ONE(a) + ONE(b-a) > ONE(b)$ iff $b_i < a_i$ for some i.

The "only if"-part of $(**)$ is clear: if $a_i \le b_i$ for all i, then
$$ONE(a) + ONE(b-a) = ONE(b).$$

Consider the "if"-part, and let i be the smallest index for which $0 = b_i < a_i = 1$. For $0 \le j \le n$, let $(b-a)_j$ be the jth bit in $b-a$, and denote $B_j = a_j + (b-a)_j$. Clearly, $B_j = b_j$ for $j < i$, whereas $b_i = 0$ and $B_i = 2$. It is easy to see that
$$\sum_{j=i+1}^{n} b_j \le 1 + \sum_{j=i+1}^{n} B_j$$
and, hence,
$$ONE(a) + ONE(b-a) = \sum_{j=0}^{n} B_j > \sum_{j=0}^{n} b_j = ONE(b).$$

This means that $(**)$ and, hence, our Claim hold true. □

Example 3.2. The interplay between b_js and B_js is illustrated by the following pairs (a, b). Recall that $a < b$. The smallest index violating dominance is $i = 0$. As is customary in binary notation, the bit a_0 is the rightmost one.

a	b	$ONE(a) + ONE(b-a)$	$ONE(b)$
000011	100000	6	1
110101	110110	5	4
100101	101010	5	3
010001	111110	6	5

□

We have now completed the auxiliary constructions needed for the proof of Theorem 3.1, and are in a position to consider a machine model and to show how computations can be given an ED representation. The machine model that we will consider is the *register machine*.

The model of a register machine that we will consider has a finite number r of *registers* $R_1, \ldots, R_r$, each capable of storing a nonnegative integer. (There is no preassigned upper bound on the size of the numbers storable in a register.) The machine is capable of adding the number 1 to the number stored in a register. Likewise, it can subtract the number 1, provided the register in question contains a positive integer. The machine is capable of testing the validity of the latter condition. Indeed, a subtraction *command* is always preceded by such a test command. The *program* of a register machine consists of a finite list of commands labelled $C_1, \ldots, C_s$. The commands are executed in the sequence given except that if C_i is a test instruction, then the next command executed is not necessarily C_{i+1}. The possible commands are *addition, subtraction*, the last command C_s being $STOP$. A subtraction command is always preceded by a *test* command which also makes transfers possible in the sequence of commands. Explicitly, the possible commands look as follows:

$$\begin{array}{ll} C_i & R_j + 1 \text{ (Add 1 to the number in } R_j), \\ C_t & IF \ R_j = 0, \ GO\ TO\ C_k, \\ C_{t+1} & ELSE \ R_j - 1 \text{ (Subtract 1 from the number in } R_j), \\ C_s & STOP. \end{array}$$

Here R_j can be any of the registers. We assume that $k \neq t+1$ in the command C_t. The computation is started from C_1 with the input x in the register R_1, the other registers containing the number 0. The input is accepted if the $STOP$ command is reached, and all registers contain 0.

A register machine is completely defined by its program. Each particular machine accepts, in the way described, a set of nonnegative integers. Acceptable sets coincide with recursively enumerable sets. (See Theorem 1.7.)

We will prove Theorem 3.1 by giving an ED representation for an arbitrary set acceptable by a register machine. We consider sets of nonnegative integers, rather than sets of n-tuples of nonnegative integers. The latter can be handled in exactly the same way except for the input format. We can either give the input in the first n registers, or else convert it to a single number by using a pairing function. We also want to emphasize that our choice of the types of commands used and the mode of acceptance is to a large extent arbitrary but it is convenient for the proof of Theorem 3.1. For instance, we could introduce other types of decisions than the one in C_t and allow other types of commands, such as a plain $GO\ TO$ command. Clearly, C_t can also be used as a $GO\ TO$ command if one register is kept empty all the time.

The next example illustrates the essential issues in the ED representation.

Example 3.3. Consider the following register machine:

Reduction to machine models

$$C_1 \quad IF \quad R_1 = 0, \quad GO\ TO \quad C_4,$$
$$C_2 \quad ELSE \quad R_1 - 1,$$
$$C_3 \quad IF \quad R_2 = 0, \quad GO\ TO \quad C_1,$$
$$C_4 \quad STOP.$$

Thus, the number r of registers equals 2, and there are $s = 4$ lines in the program. (The machine being rather "stupid" is quite irrelevant from the point of view of our discussion. It is clear that such a short program cannot accomplish much. We have omitted the $ELSE$-part of C_3 because it can never be used.)

The machine accepts the number $x = 2$ via a computation consisting of $y = 8$ steps, following the sequence of commands

$$C_1, C_2, C_3, C_1, C_2, C_3, C_1, C_4.$$

The register R_2 is empty at the entrance to each of the eight steps, whereas the register R_1 contains the number 2 at the entrance to the first two steps, the number 1 at the entrance to the next three steps, and the number 0 at the entrance to the final three steps.

The following notation is defined in terms of the parameters r, s, x, y. In our specific example the parameters assume the values 2, 4, 2, 8, respectively, but the notation is defined to fit the general case as well. Given x, we let

$$r_{i,t}, \quad 1 \le i \le r, \quad 0 \le t \le y - 1$$

be the contents of the register R_i at the entrance to the tth step of the computation, and we let

$$c_{i,t}, \quad 1 \le i \le s, \quad 0 \le t \le y - 1$$

assume the value 1 or 0 according to whether at time t we do or do not carry out the ith command in the program. (Here i refers to the label C_i, that is, the location of the command. The same command may appear at several locations.) In our specific example with $x = 2$, the r- and c-numbers corresponding to the $y = 8$ steps can be visualized by the following matrix, where time proceeds leftwards:

7	6	5	4	3	2	1	0	Time
0	0	0	1	1	1	2	2	R_1
0	0	0	0	0	0	0	0	R_2
0	1	0	0	1	0	0	1	C_1
0	0	0	1	0	0	1	0	C_2
0	0	1	0	0	1	0	0	C_3
1	0	0	0	0	0	0	0	C_4

We now start writing conditions, in ED terms, such that the conditions have a solution iff the register machine accepts x. Thus, r, s and the program are given *a priori*, and they are present in the conditions in disguised form. However, the existence of the length y of the computation is postulated. (In other words, in the definition of exponential Diophantine sets, y is among the x_i-variables and x is among the a_i-variables.)

We will express each row in the above matrix, whose entries are the numbers $r_{i,t}$ and $c_{i,t}$, as a single integer. The integer is written to base b, where b is sufficiently large, and the integers $r_{i,t}$ or $c_{i,t}$ appear as digits in the representation. More explicitly,

$$R_i = \sum_{t=0}^{y-1} r_{i,t} b^t \quad (1 \leq i \leq r),$$
$$C_i = \sum_{t=0}^{y-1} c_{i,t} b^t \quad (1 \leq i \leq s).$$

It should cause no confusion that R_i and C_i are numbers here, the interconnection with registers and commands being clear: R_i viewed as a b-ary integer encodes the history of the register R_i, and C_i similarly encodes the history of the command at the ith location. Thus, each of the digits $c_{i,t}$ is 0 or 1. Moreover, we will need the upper bound

$$r_{i,t} < b/2$$

for the digits $r_{i,t}$. It will also be convenient that b is a power of 2.

The contents of a register at time t cannot exceed $x + t$, so always $r_{i,t} \leq x + y$. So, certainly, $r_{i,t} < b/2$ if we choose

(1) $$b = 2^{x+y+s}.$$

This choice also takes care of the natural condition $s < b$.

In our specific example, $b = 2^{14} = 16\ 384$. The numbers R_1, R_2, $C_1 - C_4$ can now be computed from their representation to base B. We are not interested in the numbers themselves (they are quite large even in this overly simple case!) but, rather, in the conditions expressing the fact that the computation with input x leads to acceptance.

We need a further auxiliary number

$$U(b,y) = U = 1 + b + b^2 + \ldots + b^{y-1}.$$

Thus, all digits of U equal 1 in the b-ary representation. The number U may be defined in ED terms by the equation

(2) $$1 + bU = U + b^y.$$

Our basic auxiliary tool will be the dominance relation $\leq_D$. We use the fact that b is a power of 2. This fact implies that, whenever $\alpha \leq_D \beta$, then also each digit in the b-ary representation of α is less than or equal to the corresponding digit in the b-ary representation of β.

(For instance, take $b = 8$. If some digit of α, say the coefficient α_2 of b^2, is greater than the corresponding coefficient β_2 of β, then we cannot have $\alpha_2 \leq_D \beta_2$,

Reduction to machine models

which also leads to a violation of $\alpha \leq_D \beta$. On the other hand, every digit of α may be less than or equal to the corresponding digit of β but not $\alpha \leq_D \beta$.)

Exactly one command should be active at each time instant. This can be expressed as the conjunction of the conditions

(3) $$U = \sum_{i=1}^{s} C_i,$$

(4) $$C_i \leq_D U \quad (1 \leq i \leq s).$$

The condition

(5) $$1 \leq_D C_1$$

is needed to ensure that the program is started from the first line. We know that the *STOP* command appears only as the last command of the program, this fact being expressed in ED terms by the equation

(6) $$C_s = b^{y-1}.$$

Consider now a conditional transfer command

$$C_t \quad IF \quad R_j = 0, \quad GO\ TO \quad C_k.$$

This is taken care of by the conjunction of the conditions

(7) $$bC_t \leq_D C_k + C_{t+1},$$
(8) $$bC_t \leq_D C_{t+1} + U - 2R_j.$$

Earlier we stated the upper bound $r_{j,t} < b/2$, valid for all pairs (j,t). This bound is important for (8), and can be expressed in ED terms by

(9) $$R_j \leq_D (b/2 - 1)U \quad (1 \leq j \leq r).$$

That (7) and (8) express the conditional transfer command with the label C_t is essentially due to a simple *observation*. Assume that b is a power of 2, $b = 2^v$, and that α and β are written in b-ary notation. Assume, further, that in this notation the coefficient of b^u is odd in α but even in β. (The said coefficient may be greater in β than in α.) Then we cannot have $\alpha \leq_D \beta$ because, in binary notation, the coefficient of 2^{uv} is 0 in β but 1 in α. This observation is used in (8): the coefficient of the critical power of b in $U - 2R_j$ is odd or even according as $R_j = 0$ or $R_j > 0$ respectively.

Let us now show that (7) and (8) express our conditional transfer command with the label C_t. Recall that in each C_i every digit is either 0 or 1, the occurrence of 1 (intuitively) indicating that the command C_i is active at the corresponding time instant. The multiplication of C_i by b means a shift of one time instant: digits 1 in bC_i indicate that C_i was active at the preceding time instant. Thus, (7) means that whenever C_t was active at a certain time instant, then either C_k or C_{t+1} is activated at the next time instant. (Here the assumption $k \neq t+1$ is required; the dominance relation would not hold after the addition $C_{t+1} + C_{t+1}$.) Moreover, if $R_j = 0$ then C_{t+1} cannot be active at the next time instant because then (8) would not be satisfied. (Two 1s appearing as coefficients of b corresponding to the next time instant violate the dominance relation.) Finally, if $R_j > 0$ then (8) is only satisfied if C_{t+1} is active at the next time instant. In particular, this holds when we consider the *first* application of C_t in the course of a successful computation.

Consider the last claim in more detail. Assume that u is the first time instant when C_t is active and that $R_j > 0$. Thus, the coefficient of b^{u+1} in bC_t equals 1. Suppose C_{t+1} is not active at the time instant $u+1$. Since u is the first time instant when C_t is active and, by definition, C_{t+1} is a subtraction command that can be entered only via C_t, we conclude that the coefficient of each b^i, $i \leq u+1$, in C_{t+1} equals 0. Thus, as regards these powers of b, (8) is entirely due to the part $U - 2R_j$. The difference is positive because the register R_j contains 0 at the end of the computation. We are interested in the coefficient of b^{u+1}. Since $R_j > 0$ at time instants u and $u+1$, that is, $r_{j,u} > 0$, $r_{j,u+1} > 0$, and $2r_{j,v} < b$ holds for all v by (9), we conclude that the coefficient $b - 2r_{j,u+1}$ is even. (The inequality $2r_{j,v} < b$ also guarantees that the contents of R_j at earlier time instants do not affect the crucial coefficient.) This contradicts the dominance relation (8), since the coefficient of b^{u+1} in bC_t equals 1. This completes our argument, showing that (7) and (8) take care of the transfer command C_t.

After considering the conditional transfer commands C_t, it is now very easy to ensure that the proper command is chosen when an addition or a subtraction command C_i has been executed. The proper next command in this case is C_{i+1}, which is taken care of by the dominance relation

(10) $$bC_i \leq_D C_{i+1}.$$

Finally, we must express the condition concerning the contents of the registers, that is, concerning the addition and subtraction being carried out in the proper way. This again is easy to do, in principle in the same way as in (10), using the idea of a "time shift", caused by multiplying the numbers by b. Explicitly, we have the following equations for the numbers $R_1, \ldots, R_r$:

(11) $$R_1 = x + b(R_1 + \sum_i C_i - \sum_j C_j),$$
(12) $$R_u = b(R_u + \sum_i C_i - \sum_j C_j), \quad u = 2, \ldots, r,$$

where the i-sum is taken over all those addition commands, and the j-sum over all those subtraction commands which concern the particular register mentioned in the equation. The difference between (11) and (12) is due to the fact that, initially, R_1 contains x, whereas the other registers are empty. We know that at most one C_i or C_j is active at each time instant — at some moments a conditional transfer command may also be active. Depending on whether the active command is addition, subtraction or conditional transfer, the effect on the register in question

Reduction to machine models

is $+1$, -1 or 0 because of (11) and (12), as it should be. Since the registers are empty at the end, the first digit is 0 in each R_j and, thus, there is space for multiplication by b.

We have thus expressed the computation starting from x and consisting of y steps formally in ED terms by (1)–(12). Although we started with a specific register machine and specific values $x = 2$ and $y = 8$, we have not made any reference to this specific case in (1)–(12), and so everything also works for arbitrary register machines and computations. But let us still turn to our specific example. We have eight steps in the computation but it is convenient to speak of the time instants $0, 1, \ldots, 7$. Also, the power b^0 appears in the b-ary representations, and we want the coefficient of b^i to correspond to the ith time instant.

As regards (1), we could of course manage with a much smaller b in our specific example. In this example

$$U = 1 + b + b^2 + b^3 + b^4 + b^5 + b^6 + b^7,$$

and (2) is satisfied by $y = 8$. Relations (3)–(6) are directly inferred from the C_1–C_4-part of the matrix we gave for the description of the computation. Relation (7) assumes the form

(7)′ $\qquad bC_1 \leq_D C_4 + C_2,$
(7)″ $\qquad bC_3 \leq_D C_1 + C_4.$

Similarly, (8) assumes the form

(8)′ $\qquad bC_1 \leq_D C_2 + U - 2R_1,$
(8)″ $\qquad bC_3 \leq_D C_4 + U - 2R_2.$

Let us recall the Cs and Rs; we copy their b-ary notation from the matrix given above, leaving out the initial 0s.

$$R_1 = 11122, \quad R_2 = 0, \quad C_1 = 1001001,$$
$$C_2 = 10010, \quad C_3 = 100100, \quad C_4 = 10000000.$$

Since $U = 11111111$ in b-ary notation, and $b = 16\,384$, we infer that the b-ary digits of $u - 2R_1$ are

$$1,\ 1,\ 0,\ 16\,382,\ 16\,382,\ 16\,382,\ 16\,380,\ 16\,381.$$

Hence,

$$C_2 + U - 2R_1 = 110(16\,383)(16\,382)(16\,382)(16\,381)(16\,381).$$

Since,

$$C_4 + C_2 = 10010010,\ C_1 + C_4 = 11001001,$$
$$bC_1 = 10010010,\ bC_3 = 1001000,$$
$$C_4 + U - 2R_2 = 21111111,$$

from which (7)′, (7)″, (8)′ and (8)″ are immediately inferred.

We discuss the consequences of attempted false applications of the conditional transfer command C_1. Suppose, first, that at time instant 4 we go to C_4 rather than to C_2, that is, we make the transfer even though the register is not empty. While the validity of (7)′ is not affected, the fifth digit from the left in C_2 now being 0 causes the same digit in $C_2 + U - 2R_1$ to become 16 382. This, in turn, causes the "odd–even controversy" in the fifth digits of the two sides of (8)′ and, consequently, (8)′ cannot be valid. Suppose, secondly, that at time instant 7 we go to C_2 rather than to C_4, that is, we do not make the transfer even though the register is empty. Now the eighth digit from the left is 1 in C_2 but 0 in C_4. Since $R_1 = 00011122$ and, thus, the eighth digit from the left in $C_2 + U - 2R_1$ equals 2, we have the "odd–even controversy" as regards this digit in (8)′.

We still consider the conditions (9)–(12) in our specific example. Since $b/2 - 1 = 8191$, it is clear that (9) is satisfied. (10) now assumes the form $bC_2 \leq_D C_3$, which holds true. Consider, finally, (11) and (12). There are no addition commands, and C_2 is the only subtraction command, R_1 being the register involved. Thus, both of the sums in (12) are empty, and (12) assumes the simple form $R_2 = 0 = bR_2$. Since (in b-ary notation)

$$R_1 - C_2 = 11122 - 10010 = 1112 ,$$

Equation (11) becomes the valid b-ary equation in this specific case

$$11122 = 2 + b \cdot 1112 .$$

□

We have discussed the specifics of Example 3.3, a particular register machine with a particular input. We have shown how to represent a successful computation in ED terms. Our discussion has been very lengthy because we have presented everything in such a way that it also applies for a general register machine and input. In this way we have also given a proof for the Davis–Putnam–Robinson Theorem, stated here as Theorem 3.1. Let us recapitulate, begin everything from the beginning and see how the construction works.

We are given a recursively enumerable set S of nonnegative integers. We want to show that S is exponential Diophantine, that is, to give "an ED representation" for S. This, in turn, means the construction of two functions P and Q, built from the variables $x, y_1, \ldots, y_m$ and nonnegative integer constants by the operations of addition $A + B$, multiplication AB and exponentiation A^B, such that x is in S iff, for some $y_1, \ldots, y_m$, $P(x, y_1, \ldots, y_m) = Q(x, y_1, \ldots, y_m)$. For the purposes of Chapter 5, we observe, at the same time, that the m-tuple $(y_1, \ldots, y_m)$ is unique. In other words, the ED representation obtained is *singlefold*.

We know that S is accepted by a register machine M. Let r be the number of registers and s the number of commands in the program defining M. The types of commands in the program, as well as the input–output format of M, follow the

Reduction to machine models

conventions specified in the above discussion. Consequently, a number x is in S iff the machine M, when started at the first command C_1 with x in register R_1 and with other registers empty, reaches the last command C_s ($STOP$) with all registers empty. If this happens, the computation has a length y uniquely determined by x (because M is deterministic). Of course, there is in general no recursive upper bound for y in terms of x; whenever such a bound exists, S is recursive.

Our ED representation for S is based on M: we use the fact that S is accepted by M. In our discussion we will not write down explicit functions P and Q as given in the definition. Of course, it is in principle possible to construct P and Q explicitly from the proof. We use the facts about ED constructions developed at the beginning of Section 3.1. Of these facts, Lemma 3.4 concerning the dominance relation is the most important one.

We are now ready to present the essence of the proof of Theorem 3.1 in the form of a very compact statement, referred to as the *exponential Diophantine characterization of recursive enumerability via register machines*, or, briefly, *ED char RE via RM*. We are given a recursively enumerable set S, accepted by the register machine M with r registers and s commands in the program. We denote by T the set of indices t such that the program contains the conditional transfer command

$$C_t \ IF \ R_{j(t)} = 0 \ , \ GO \ TO \ C_{k(t)} \ .$$

(Here we use the notation $j(t)$ and $k(t)$, instead of our earlier j and k, to emphasize dependence on t.) Similarly, A is the set of indices i such that C_i is an addition or a subtraction command in the program. Finally, for each $j = 1, \ldots, r$, $ADD(j)$ (resp. $SUB(j)$) is the set of indices i such that C_i is an addition (resp. a subtraction) command concerning the register R_j.

ED char RE via RM. A nonnegative integer x is in S iff, for some unique nonnegative integers $y, b, U, R_1, \ldots, R_r, C_1, \ldots, C_s$, the following conditions (1)–(12) are satisfied. Uniqueness follows because M is deterministic and shows that the ED representation is singlefold. This will be important in Chapter 5.

(1) $$b = 2^{x+y+s} ,$$

(2) $$1 + bU = U + b^y ,$$

(3) $$U = \sum_{i=1}^{s} C_i ,$$

(4) $$C_i \leq_D U \quad (i = 1, \ldots, s) ,$$

(5) $$1 \leq_D C_1 ,$$

(6) $$C_s = b^{y-1} ,$$

106   Chapter 3. Diophantine Problems

(7) $$bC_t \leq_D C_{k(t)} + C_{t+1} \quad (t \in T),$$

(8) $$bC_t \leq_D C_{t+1} + U - 2R_{j(t)} \quad (t \in T),$$

(9) $$R_j \leq_D (b/2 - 1)U \quad (j = 1, \ldots, r),$$

(10) $$bC_i \leq_D C_{i+1} \quad (i \in A),$$

(11) $$R_1 = x + b(R_1 + \sum_{i \in ADD(1)} C_i - \sum_{i \in SUB(1)} C_i),$$

(12) $$R_u = b(R_u + \sum_{i \in ADD(u)} C_i - \sum_{i \in SUB(u)} C_i) \quad (u = 2, \ldots, r).$$

□

That our *ED char RE via RM* shows that S is indeed exponential Diophantine follows by Lemma 3.4 and the fact that the conjunction of two *ED* relations is itself *ED*. (Recall that the latter fact follows because $P = 0$ and $Q = 0$ iff $P^2 + Q^2 = 0$.) Considering the *ED* representation resulting from Lemma 3.4, it is indeed within the reach of reasonable computations to present (1)–(12) in terms of a single equation, as required in the definition. However, we do not undertake this task here.

Instead, we say a few words about the proof of *ED char RE via RM*, although the details should be fairly clear from the discussion in Example 3.3. In particular, we have shown that if there is a successful computation of M with input x, then (1)–(12) hold when y is the length of the computation and the Rs and Cs are associated with the registers and commands in the way described above. What about the converse? Let us examine this matter.

Suppose that (1)–(12) hold for some numbers $y, b, U, R_1, \ldots, R_r, C_1, \ldots, C_s$. Thus, b is a power of 2 and, by (2),

$$U = 1 + b + b^2 + \ldots + b^{y-1}.$$

Relations (3) and (4) show that the b-ary representation of every C_i consists of 0s and 1s and, moreover, that for each $i = 0, \ldots, y - 1$, there is exactly one C_j such that the ith digit in C_j equals 1. Hence, we may associate with the Cs the intuitive meaning: the command C_j is the active one at the ith time instant. We then know that two commands are never active at the same time instant. On the other hand, it is possible that some C_j equals 0. The intuitive meaning is that the jth command is never used in the computation with input x.

With this intuitive meaning of the numbers C_j in mind, (5) and (6) show that the first and last commands in the computation are the proper ones. We have shown in detail how (7)–(10) guarantee that the next command chosen is always the proper one. The intuitive meaning associated with the numbers R_j is the one described above: for $i = 0, \ldots, y-1$, the ith digit from the left in R_j is the number in the jth register at the ith time instant. (Thus, C_j describes the history of the application of the jth command, and R_j the history of the contents of the jth register.) For technical reasons due to (8), we do not want the contents of any register to grow too large — this is taken care of by (9). We have also shown in detail how (11) and (12) guarantee that the contents of the registers are changed according to the commands; in this statement we again have the intuitive meaning of the Cs and Rs in mind. We have also seen that all registers become empty at the end.

We have established our *ED char RE via RM* and, hence, also Theorem 3.1.
□

It is interesting to observe that input x appears only in (1) and (11). It must be in (11) — x is in the first register at the beginning. The appearance in (1) could be avoided because if x is large, so also is y.

The approach of this section is by no means restricted to our particular definition of a register machine. Other types of machine models (other kinds of conditional transfer commands, other kinds of input–output formats) can be handled by similar techniques.

3.2 Hilbert's Tenth Problem

Theorem 3.1 can still be essentially strengthened: every recursively enumerable set and relation is Diophantine. This shows that Hilbert's Tenth Problem is undecidable: an algorithm for solving Hilbert's Tenth Problem would yield an algorithm for solving the membership problem for recursively enumerable sets.

Clearly, the set of solutions for a given Diophantine (polynomial) equation is recursively enumerable, and so is the set of all equations possessing solutions. Thus, here again we have an effective procedure for listing all elements of a set (all solutions of a given equation or all solvable equations) but we cannot decide the emptiness of the set, since there is no way of knowing when to stop the listing procedure if no elements have been produced. Indeed, drastic examples are known, showing that in some cases we have to wait for quite a long time. In the smallest positive solution (x, y) of the Diophantine equation

$$x^2 = 991y^2 + 1$$

x has 30 and y 29 decimal digits.

Examples of Diophantine relations were given in Lemma 3.2, where exponentiation was not used at all. The set S_{comp} of composite numbers is Diophantine because

$$x \in S_{comp} \text{ iff } (\exists y)(\exists z)(x = (y+2)(z+2)).$$

Being recursively enumerable, the set of prime numbers is Diophantine but it obviously cannot be represented simply. We have defined Diophantine sets as sets for which a certain polynomial equation possesses a solution in nonnegative integers. It is often easier to deal with sets of values of a polynomial. Therefore, the following result is pleasing. We restrict our attention to sets consisting of positive (rather than nonnegative) integers.

Theorem 3.5. For every Diophantine set S of positive integers, there is a polynomial Q such that S equals the set of positive values of Q when the variables of Q range over nonnegative integers.

Proof. We know that there is a polynomial P such that

$$x \in S \text{ iff } (\exists y_1 \geq 0)\ldots(\exists y_m \geq 0)(P(x, y_1, \ldots, y_m) = 0) \, .$$

Define the polynomial Q by

$$Q(x, y_1, \ldots, y_m) = x(1 - (P(x, y_1, \ldots, y_m))^2) \, .$$

Clearly, every element of S is in the range of Q. Conversely, a positive integer x can be in the range of Q only if there are nonnegative integers $y_1, \ldots, y_m$ with the property $P(x, y_1, \ldots, y_m) = 0$.

□

As an illustration without any explanations we give here a polynomial Q such that the set of positive values of Q equals the set of prime numbers. The polynomial uses the letters of the English alphabet. The reader is referred to [Sal 2] for discussion and bibliographical references concerning this polynomial and the "Fermat polynomial" presented below:

$$\begin{aligned}
Q(a,\ldots,z) = \ & (k+2)\{1 - [wz + h + j - q]^2 \\
& - [(gk + 2g + k + 1)(h + j) + h - z]^2 \\
& - [2n + p + q + z - e]^2 - [16(k+1)^3(k+2)(n+1)^2 + 1 - f^2]^2 \\
& - [e^3(e+2)(a+1)^2 + 1 - o^2]^2 - [(a^2 - 1)y^2 + 1 - x^2]^2 \\
& - [16r^2 y^4(a^2 - 1) + 1 - u^2]^2 \\
& - [((a + u^2(u^2 - a))^2 - 1)(n + 4dy)^2 + 1 - (x + cu)^2]^2 \\
& - [n + 1 + v - y]^2 \\
& - [(a^2 - 1)l^2 + 1 - m^2]^2 - [ai + k + 1 - l - i]^2 \\
& - [p + l(a - n - 1) + b(2an + 2a - n^2 - 2n - 2) - m]^2 \\
& - [q + y(a - p - 1) + s(2ap + 2a - p^2 - 2p - 2) - x]^2 \\
& - [z + pl(a - p) + t(2ap - p^2 - 1) - pm]^2\} \, .
\end{aligned}$$

Thus, questions about prime numbers can be formulated as questions about positive values of the polynomial Q. For instance, the celebrated open problem

Hilbert's Tenth Problem

about the existence of infinitely many twin primes can be stated as follows: does Q have infinitely many positive values α such that $\alpha - 2$ is also a value of Q?

Diophantine sets S are defined by a condition of the form

(*) $\qquad x \in S$ iff $(\exists y_1 \geq 0)\ldots(\exists y_m \geq 0)(P(x, y_1, \ldots, y_m) = 0)$,

where P is a polynomial with integer (possibly negative) coefficients. Given S, the polynomial P is by no means unique. For a fixed S, let us refer to the least degree of P for which (*) is satisfied as the *degree* of S and the least value of m as the *dimension* of S. In general, the degree and dimension are not reached by the same polynomial because if one tries to reduce the degree of the polynomial, then one usually has to increase the number of variables, and vice versa. The best-known combined estimate is the pair $(5,5)$: every Diophantine set S has a representation (*), where $m \leq 5$ and P is of degree ≤ 5. As regards the degree alone, one can do better.

Theorem 3.6. *Every Diophantine set is of degree less than or equal to 4.*

Proof. We begin with a representation (*) for a Diophantine set S. We introduce new variables z_j, and replace the equation $P = 0$ by simultaneous quadratic equations. More explicitly, each term $a\alpha_i x_j$ of degree ≥ 3 in P is first replaced by $az_i x_j$, where z_i is a new variable and a is the coefficient of the original term. In this fashion the equation $P = 0$ becomes a quadratic equation $P' = 0$. This quadratic equation is combined with all equations $z_i = \alpha_i$, resulting from the different terms in P. The system of simultaneous equations is of degree less than the degree of P. The procedure is then applied to all equations $z_i = \alpha_i$ having a degree at least 3 until, finally, a system of simultaneous quadratic equations

$$P_1 = 0, \ldots, P_n = 0$$

is obtained. The system is then replaced by the single equation $P_1^2 + \cdots + P_n^2 = 0$ of degree at most 4. $\square$

Example 3.4. Consider the Diophantine set S defined by

$$x \in S \text{ iff } (\exists y_1 \geq 0)(\exists y_2 \geq 0)(x^3 y_1 y_2^2 - 3xy_1 y_2 - 2 = 0).$$

The equation involved is first replaced by the system of equations

$$z_1 y_2 - 3z_2 y_2 - 2 = 0,$$
$$z_2 = xy_1, \ z_1 = z_3 y_2, \ z_3 = z_4 y_1,$$
$$z_4 = z_5 x, \ z_5 = x^2,$$

whence the original definition may be replaced by

$$x \in S \text{ iff } (\exists y_1, y_2, z_1, \ldots, z_5 \geq 0)[(z_1 y_2 - 3z_2 y_2 - 2)^2$$
$$+ (xy_1 - z_2)^2 + (y_2 z_3 - z_1)^2 + (z_4 y_1 - z_3)^2$$

$$+(z_5 x - z_4)^2 + (x^2 - z_5)^2 = 0] \,.$$

□

We now state formally the most important result in this chapter.

Theorem 3.7. All recursively enumerable sets and relations are Diophantine.

Theorem 3.7 is an immediate corollary of Theorem 3.1 and the following result, referred to as the *Matijasevitch Theorem*.

Theorem 3.8. The ternary relation $m = n^k$ is Diophantine.

While Hilbert's Tenth Problem dates back to 1900, Theorem 3.1 is roughly from 1960 and Theorem 3.8 from 1970. The proof of Theorem 3.8 will be given at the end of this section. Some details of the proof are purely number-theoretic and, as such, are not of any major concern to us in this book. They will be stated as lemmas without proofs. An interested reader will find many detailed proofs in the literature. (See [Bör] and the references given there.)

We now deduce some consequences of Theorem 3.7. Let S_i, $i = 1, 2, \ldots$, be an enumeration of all recursively enumerable sets of nonnegative integers. Because the binary relation $x \in S_i$ is recursively enumerable, the following result is an immediate corollary.

Theorem 3.9. There is a polynomial P_u with integer coefficients such that, for all i, the ith recursively enumerable set S_i satisfies the condition

$$x \in S_i \text{ iff } (\exists y_1 \geq 0) \ldots (\exists y_m)[P_u(i, x, y_1, \ldots, y_m) = 0]$$

for every x.

The polynomial P_u is referred to as a *universal polynomial*: a Diophantine representation for every recursively enumerable set is obtained in terms of this single polynomial by letting the first variable range over positive integers. Similarly, one may also construct a universal polynomial P'_u corresponding to the modification presented in Theorem 3.5. By fixing the value of one variable in P'_u, we obtain a representation of an arbitrary recursively enumerable set as the set of positive values of P'_u.

The polynomial P_u of Theorem 3.9 can be used to obtain the following stronger version for the result about the undecidability of Hilbert's Tenth Problem. This stronger version is immediate because the emptiness problem is undecidable for recursively enumerable sets.

Theorem 3.10. There is a polynomial $P(x, y_1, \ldots, y_m)$ with integer coefficients such that no algorithm exists for deciding whether or not an arbitrary equation of the form

$$P(x_0, y_1, \ldots, y_m) = 0 \,,$$

where x_0 is a positive integer, has a solution in nonnegative integers $y_1, \ldots, y_m$.

Hilbert's Tenth Problem

In the next chapter we will discuss the borderlines between decidability and undecidability. A natural parameter in this respect for polynomial Diophantine equations is the degree of the polynomial. It can be shown that the solvability of Diophantine equations with degree ≤ 2 is decidable. We can infer by Theorem 3.6 that this problem is undecidable for equations of degree 4. For degree 3 the problem is open.

By Theorem 3.7, many famous open problems can be expressed as problems concerning positive values of polynomials. We do this for Fermat's celebrated last problem, that is, we give a Diophantine representation for the quaternary relation

$$p_1^k + p_2^k = p_3^k \, .$$

For simplicity, we restrict our attention to the positive values of the variables:

$$p_1^k + p_2^k = p_3^k \quad \text{iff} \quad (\exists q_1, q_2, q_3)(q_1 = p_1^k \text{ and } q_2 = p_2^k \text{ and } q_3 = p_3^k \text{ and } q_1 + q_2 = q_3)$$
$$\text{iff} \quad (\exists a_1, a_2, a_3, b_1, b_2, b_3, \ldots, \tau_1, \tau_2, \tau_3)$$

$$\{\sum_{i=1}^{3} [(a_i - q_i - p_i - k - p_i^2 - 2)^2 + (u_i^2 - a_i(a_i + 2)u_i w_i + a_i^2 w_i^2 - 1)^2$$
$$+ (l_i - u_i - b_i)^2 + (l_i - v_i - c_i)^2 + (l_i^2 - l_i z_i - z_i^2 - 1)^2$$
$$+ (g_i^2 - g_i h_i - h_i^2 - 1)^2 + (g_i - d_i l_i^2)^2 + (m_i - 2 - e_i l_i)^2$$
$$+ (m_i - 3 - j_i(2h_i + g_i))^2 + (x_i^2 - m_i x_i y_i + y_i^2 - 1)^2$$
$$+ (x_i - u_i - n_i l_i)^2 + (x_i - v_i - r_i(2h_i + g_i))^2 + (s_i^2 - u_i s_i t_i + t_i^2 - 1)^2$$
$$+ (t_i - s_i - \theta_i)^2 (s_i - t_i)^2 + (v_i - 9t_i - \sigma_i)^2 + (s_i - k - \rho_i(u_i - 2))^2$$
$$+ (t_i + (p_i - u_i)s_i - q_i - \tau_i(p_i u_i - p_i^2 - 1))^2] + (q_1 + q_2 - q_3)^2 = 0\} \, .$$

The remainder of this section is devoted to the proof of Theorem 3.8. We will omit the proof of certain purely number-theoretic facts, presented as Lemmas A–E.

The number-theoretic background concerns numbers of the form $x + y\sqrt{\alpha}$, where x and y are nonnegative integers and α is a fixed positive integer that is not a square. The sum and product of two numbers of this form is again a number in the same form. Moreover, the representation of a number in this form is unique: the equation $x_1 + y_1\sqrt{\alpha} = x_2 + y_2\sqrt{\alpha}$ is possible only if $x_1 = x_2$ and $y_1 = y_2$.

Let $a > 1$ be an integer. Then $a^2 - 1$ is not a square; we choose $\alpha = a^2 - 1$. By the uniqueness of the representation, for each $k \geq 0$, there is exactly one pair (x, y) of nonnegative integers such that

$$x + y\sqrt{a^2 - 1} = (a + \sqrt{a^2 - 1})^k \, .$$

For small values of k, the pairs are as follows.

$$\begin{aligned}
k = 0 &: \quad x = 1, \ y = 0, \\
k = 1 &: \quad x = a, \ y = 1, \\
k = 2 &: \quad x = 2a^2 - 1, \ y = 2a, \\
k = 3 &: \quad x = 4a^3 - 3a, \ y = 4a^2 - 1, \\
k = 4 &: \quad x = 8a^4 - 8a^2 + 1, \ y = 8a^3 - 4a \, .
\end{aligned}$$

For an arbitrary k, we use the notation

$$x = X(k, a), \ y = Y(k, a).$$

Thus, $X(4, a) = 8a^4 - 8a^2 + 1$ and $Y(4, a) = 8a^3 - 4a$.

It turns out that the numbers $X(k, a)$ and $Y(k, a)$ constitute exactly the nonnegative integer solutions of the *Pell-equation* with the *parameter a*:

$$x^2 - (a^2 - 1)y^2 = 1.$$

Lemma A. Let $a > 1$ be fixed. Then every pair $(X(k, a), Y(k, a))$ is a nonnegative integer solution of the Pell-equation and, conversely, for every solution (x, y) with $x \geq 0$, $y \geq 0$, there is a k such that $x = X(k, a)$ and $y = Y(k, a)$.

Our aim is to present a Diophantine representation of the exponential relation of Theorem 3.8 in terms of the ternary relation $x = X(k, a)$ and then show that this relation is also Diophantine. More explicitly, our proof can be separated into the proof of two subgoals.

Subgoal 1. The relation $m = n^k$ can be given a Diophantine representation in terms of the relation $x = X(k, a)$, that is, if the latter relation is Diophantine, so is the former relation.

Subgoal 2. The ternary relation $x = X(k, a)$ is Diophantine.

Clearly, Theorem 3.8 is a direct consequence of Subgoals 1 and 2. In the proof of Subgoal 1 we need three number-theoretic lemmas. The very simple initial observation is that it is sufficient to prove that m and n^k are congruent with respect to a certain modulus and are smaller than the modulus; then the equation $m = n^k$ follows. The difficulty lies in expressing the matter in Diophantine terms. The next lemma can be used to show that $2an - n^2 - 1$ is a suitable modulus.

Lemma B. For all $a > 1$ and all positive k and n,

$$X(k, a) - Y(k, a)(a - n) \equiv n^k \pmod{2an - n^2 - 1}.$$

Example 3.5. Let us verify the lemma for $k = 3$. We noticed above that

$$X(3, a) = 4a^3 - 3a, \ Y(3, a) = 4a^2 - 1.$$

Hence,

$$X(3, a) - Y(3, a)(a - n) = (4a^2 - 1)n - 2a.$$

The claim now follows because

$$(4a^2 - 1)n - 2a - n^3 = (2an - n^2 - 1)(n + 2a).$$

□

By Lemma B, the four conditions

(1) $$x^2 - (a^2 - 1)y^2 = 1,$$

(2) $$x = X(k, a),$$

(3) $$x - y(a - n) \equiv m \pmod{2an - n^2 - 1},$$

and

(4) $$m < 2an - n^2 - 1$$

imply the equation $m = n^k$, provided

(*) $$n^k < 2an - n^2 - 1.$$

Our task is to express (*) in Diophantine terms. In fact, (*) is a consequence of the inequality

(*)' $$n^k < a.$$

If $n = 1$, then both (*)' and (*) hold; recall that $a > 1$, which condition will also be postulated in (5) below. If $1 < n \leq n^k < a$, we also obtain

$$a < (na - 1) + n(a - n) = 2an - n^2 - 1$$

and, hence, (*)' implies (*).

To express (*)' in Diophantine terms, we need facts concerning the growth of the solutions of the Pell-equation. The growth is exponential in the sense made explicit in Lemma C below. This opens the possibility of applying the following technique that can be viewed as *the key idea in the proof*. The parameter a of the original Pell-equation is expressed as the X-component of the solution of *another Pell-equation*. In this way, exponential growth is reduced to polynomial expressions. In Lemma C the parameter of the Pell-equation (> 1) is denoted by b.

Lemma C. $b^i \leq X(i, b)$, $X(i, b) < X(i + 1, b)$ and $Y(i, b) \equiv i \pmod{b - 1}$.

Thus, in addition to growth properties, Lemma C also provides a congruence between the Y-solution and its solution number. To obtain the inequality (*)', we now apply Lemma C in the following fashion. The aim is to present the parameter a in the form

(*)'' $$a = X(i, b),$$

where the new parameter b exceeds both n and k and the solution number i satisfies the inequality

(*)''' $$b - 1 \leq i.$$

Suppose we have found i and b sufficiently large such that $(*)''$ and $(*)'''$ hold. By the inequalities of Lemma C,
$$n^k < b^k \leq b^{b-1} \leq X(b-1,b) \leq X(i,b) = a$$
and, hence, $(*)'$ holds. So it suffices to construct i and b satisfying $(*)''$ and $(*)'''$.

Let us first postulate the necessary inequalities:

(5) $\qquad\qquad 1 < a\,,\ 1 < b\,,\ n < b\,,\ k < b\,.$

Then consider the congruence in Lemma C. To satisfy $(*)''$, we want a to be the X-component of the ith solution of the Pell-equation with parameter b, that is, a satisfies the equation
$$a^2 - (b^2 - 1)y_1^2 = 1\,.$$
By the congruence in Lemma C, $y_1 \equiv i \pmod{b-1}$. If we now postulate that y_1 is divisible by $b-1$, that is, $y_1 = (b-1)z$, then i must also be divisible by $b-1$. Since $i = 0$ is not possible (then a would be the X-component of the 0th solution, implying $a = 1$), we conclude that $(*)'''$ holds. Thus, the equation we have to postulate can be written as

(6) $\qquad\qquad a^2 - (b^2 - 1)((b-1)z)^2 = 1\,.$

We now have all the machinery needed for Subgoal 1. We *claim* that $m = n^k$ holds for positive m, n, k iff there are nonnegative a, b, x, y, z such that each of the conditions (1)–(6) is satisfied. By Lemma 3.2, our claim brings us to Subgoal 1. (Indeed, apart from (2), all conditions are polynomial equations, inequalities and congruences.)

To prove the claim, assume first that (1)–(6) are satisfied. Equation (1) shows that (x,y) constitutes a solution of the Pell-equation with parameter a. Hence, by (2) and Lemma A, $y = Y(k,a)$. By (3) and Lemma B,

$(**)$ $\qquad\qquad m \equiv n^k \pmod{2an - n^2 - 1}\,.$

By (6) and Lemma A, we can write $a = X(i,b)$, $(b-1)z = Y(i,b)$, for some i. Although i is unknown to us, the congruence in Lemma C tells us that $(*)'''$ holds. Hence, we can use (5) and the inequalities in Lemma C to obtain the inequality $(*)'$, from which $(*)$ follows, $(*)$. Relations (4) and $(**)$ finally give the result $m = n^k$.

Conversely, assume that $m = n^k$ holds. We want to show that (1)–(6) hold, for some a, b, x, y, z. Choose a large enough b such that the last three inequalities in (5) are satisfied. Define $a = X(b-1, b)$. Then the first inequality in (5) is also satisfied and, moreover, by Lemma C,
$$n^k < b^k \leq b^{b-1} \leq X(b-1, b) = a\,.$$
Thus, $(*)'$ holds, implying that $(*)$ holds. Since $m = n^k$, we obtain (4). We now choose x to satisfy (2), and define $y = Y(k, a)$. By Lemma A, (1) is satisfied and,

Hilbert's Tenth Problem

by Lemma B and the hypothesis $m = n^k$, (3) is also satisfied. Since $a = X(b-1, b)$, we obtain by Lemma A,

$$a^2 - (b^2 - 1)(Y(b-1, b))^2 = 1.$$

Because, by Lemma C,

$$Y(b-1, b) \equiv b - 1 \equiv 0 \pmod{b - 1},$$

we may write $Y(b-1, b) = (b-1)z$. Consequently, for this z, (6) holds. We have established our claim and thus reached Subgoal 1. We still have to prove that the ternary relation $x = X(k, a)$ is Diophantine, in order to reach Subgoal 2 also.

We again postulate a sequence of conditions, expressed in Diophantine terms and also involving variables other than x, k, a, and show that $x = X(k, a)$ holds iff the conditions are satisfied for some values of the additional variables. We begin again with the condition

(1) $$x^2 - (a^2 - 1)Y^2 = 1,$$

expressing the fact that x is the X-component of a solution of the Pell-equation with parameter a. By Lemma A, we know that $x = X(i, a)$ and $y = Y(i, a)$, for some i. We will impose additional conditions implying that the solution number i is, in fact, equal to k. In these conditions other Pell-equations will appear. In this sense the construction resembles that used in the proof of Subgoal 1. To assist the reader, we give the following explanations concerning the notation in advance. The variables x, k, a are those appearing in our ternary relation, whereas c, y, s, t, u, v are additional variables used in the Diophantine representation, given by conditions (1)–(9). Conditions marked by stars, as well as the variables i, j, n only appear in our argument concerning the Diophantine representation.

We will use another solution $u = X(n, a)$, $v = Y(n, a)$ of the Pell-equation with the parameter a,

(2) $$u^2 - (a^2 - 1)v^2 = 1,$$

as well as the Pell-equation with another parameter c,

(3) $$s^2 - (c^2 - 1)t^2 = 1.$$

Let (s, t) be the jth solution, $s = X(j, c)$, $t = Y(j, c)$. Important properties of the "hidden variables"" i, j, n are implied by certain conditions imposed on the other variables. These properties will imply that $k = i$.

We begin with the condition

(4) $$k \equiv t \pmod{4y}.$$

Since, by Lemma C, $t = Y(j, c) \equiv j \pmod{c - 1}$, we obtain the congruence

(∗) $$k \equiv j \pmod{4y},$$

provided that we impose the additional condition

(5) $\qquad 4y|(c-1)\,,\ c-1>1\,.$

Still using other conditions, we will prove that

$(*)'\qquad\qquad\qquad j\equiv i\ (\mathrm{mod}\ 4y)\,,$

from which the important congruence

$(**)\qquad\qquad\qquad k\equiv i\ (\mathrm{mod}\ 4y)$

follows by $(*)$.

We need the following lemma about the properties of the solutions of the Pell-equation. The lemma ties divisibility properties of solutions together with divisibility properties of solution numbers. The parameters a and c are assumed to be >1, and $0 < i \leq n$.

Lemma D. (i) If $X(j,a) \equiv X(i,a)\ (\mathrm{mod}\ X(n,a))$, then $j \equiv i\ (\mathrm{mod}\ 4n)$.

(ii) If $a \equiv c\ (\mathrm{mod}\ d)$, then $X(j,a) \equiv X(j,c)\ (\mathrm{mod}\ d)$.

(iii) If $(Y(i,a))^2 | Y(n,a)$, then $Y(i,a)|n$.

Recall that $X(n,a) = u$. Thus $(*)'$ follows by Lemma D(i) if the following two conditions are satisfied:

$(*)_1 \qquad\qquad\qquad X(j,a) \equiv X(i,a)\ (\mathrm{mod}\ u)\,,$

$(*)_2 \qquad\qquad\qquad y|n,\ 0 < i \leq n\,.$

Since $Y(i,a) = y$ and $Y(n,a) = v$, $(*)_2$ follows by Lemma D(iii) if we impose the condition

(6) $\qquad\qquad\qquad y^2 | v\,.$

By Lemma D(iii), (6) implies $y|n$. Relation (6) also implies that $y \leq v$, from which $i \leq n$ follows by monotonic growth of the Y-components. If we impose

(7) $\qquad\qquad\qquad a \equiv c\ (\mathrm{mod}\ u)\,,$

we obtain $X(j,a) \equiv s\ (\mathrm{mod}\ u)$, by Lemma D(ii). From this $(*)_1$ follows if we further impose

(8) $\qquad\qquad\qquad s \equiv x\ (\mathrm{mod}\ u)\,.$

Since we now have both $(*)_1$ and $(*)_2$, we conclude that $(*)'$ and, consequently, $(**)$ also follows from the conditions imposed so far. Since, clearly, the solution number does not exceed the Y-component of the solution, we obtain $i \leq y$. If we impose the further condition

(9) $$k \leq y,$$

we obtain by (∗∗) the equation $k = i$ and, hence, $x = X(k, a)$.

We now *claim* that $x = X(k, a)$ holds for some triple (x, k, a) with $a > 1$ iff conditions (1)–(9) are satisfied for some c, y, s, t, u, v. Clearly, this claim proves Subgoal 2.

In fact, we have already established the "if"-part of the claim. If (1)–(9) are satisfied we first write $x = X(i, a)$ by (1), and then conclude by the argument given above that $i = k$ follows by (1)–(9). Conversely, assume that $x = X(k, a)$ with $a > 1$. We show how we can choose the values for the variables in such a way that (1)–(9) hold.

We first choose $y = Y(k, a)$; then (1) will be satisfied. The choice

$$u = X(2kY(k,a), a), \ v = Y(2kY(k,a), a)$$

clearly satisfies (2). We still need some divisibility properties of the solutions of the Pell-equations. In the statement of Lemma E, $a > 1$ is again assumed.

Lemma E. (i) $X(n, a)$ and $Y(n, a)$ are relatively prime. (ii) The solution number n is even iff $Y(n, a)$ is even. (iii) $(Y(k,a))^2$ divides $Y(2kY(k,a), a)$.

Lemma E(iii) is the same as (6) in our present notation. (In fact, E(iii) can be stated in a somewhat stronger form but we have stated it tailor-made for our purposes.) By Lemma E(ii), v is even. Hence, by Lemma E(i), u is odd. This means that any prime factor of u dividing $4y$ would have to divide y and hence, by (6), also v, which would contradict Lemma E(i). Consequently, u and $4y$ are relatively prime and thus the simultaneous congruences

$$c \equiv a \pmod{u}, \ c \equiv 1 \pmod{4y}$$

have a solution $c > 1$. This solution will be our choice for c. When we define

$$s = X(k, c), \ t = Y(k, c),$$

we have fixed the values of all variables appearing in (1)–(9).

Equation (3) is satisfied by the choice of s and t. The congruence $c \equiv 1 \pmod{4y}$ guarantees that (5) is satisfied, and c was directly defined to satisfy (7). By Lemma D(ii),

$$s = X(k, c) \equiv X(k, a) = x \pmod{u},$$

which gives (8). Relation (9) holds because the solution number does not exceed the Y-component. By Lemma C, we obtain

$$t = Y(k, c) \equiv k \pmod{c-1},$$

from which (4) follows by (5). Thus, (1)–(9) hold, and we have established our claim. We have also reached Subgoal 2 and, hence, have completed the proofs of Theorems 7 and 8.

3.3 Power series and plant development

Because of its very fundamental arithmetical nature, Hilbert's Tenth Problem is a very suitable point of reference in reduction arguments concerning problems of various types. In the same fashion as in Section 2.4 for the Post Correspondence Problem, we now give some typical undecidability proofs based on Hilbert's Tenth Problem. The purpose is simply to give an idea of some reductions rather than to present any formal grounds concerning the suitability of specific types of problems for specific types of reductions.

In the problem areas considered, auxiliary tools have to be developed before reduction techniques can be applied. We will present the details of such developments only to the extent that they illustrate the connection with Diophantine representations.

Formal power series in noncommuting variables constitute a powerful tool in various problem areas, ranging from algebra and combinatorics to formal languages and developmental models in biology. The series being formal means, roughly speaking, that we are not interested in summing up the series but, rather, in various operations on the series. That the variables do not necessarily commute is quite essential for applications, where the order of letters is important. For our purposes it suffices to present a somewhat restricted version of the basic notions.

A *formal power series* r is a mapping of the set V^* of all words over an alphabet V into the set Z of integers. It is customary to denote by (r, w) the value of the mapping r for the argument $w \in V^*$, and to present the mapping itself as a formal sum

$$r = \sum_{w \in V^*} (r, w) w \ .$$

Then the *sum*, *difference* and *product* of two series r and r' can be defined in the following (intuitively very plausible) way:

$$\begin{aligned} r + r' &= \sum_{w \in V^*} ((r, w) + (r', w)) w \ , \\ r - r' &= \sum_{w \in V^*} ((r, w) - (r', w)) w \ , \\ rr' &= \sum_{w \in V^*} (\sum_{uv = w} (r, u)(r', v)) w \ . \end{aligned}$$

By comparison with ordinary multiplication, the only difference is that the order of the letters of w cannot be changed. Another operation needed in the following is the term-wise or *Hadamard product*, defined by

$$r \odot r' = \sum_{w \in V^*} (r, w)(r', w) w \ .$$

Power series and plant development

The values (r, w) are referred to as the *coefficients* of the series r. Coefficients 1 are omitted, and so are terms having the coefficient 0. The language $\{w \in V^* | (r, w) \neq 0\}$ is called the *support* of the series r. Many questions concerning formal languages can be answered by studying the supports of formal power series. Given a language L, the support of its *characteristic series* $\text{char}(L) = \sum_{w \in L} w$ equals the language itself. Thus, all nonzero coefficients in the characteristic series equal 1.

In our discussion the coefficients will be integers. The basic setup can be more general, and the coefficients may come from an arbitrary semiring. The term "Z-rational" defined below reflects our more restricted setup.

A formal power series r is called *Z-rational* iff the coefficients are obtained in the following special way. For some $n \geq 1$, there is a morphism h of V^* into the multiplicative monoid of n-dimensional square matrices with integer entries, such that the coefficient (r, w) equals the upper right-hand corner entry of $h(w)$, for all w.

Thus, to define a Z-rational series, one has to specify square matrices M_x, all of the same dimension, for all letters x of the alphabet V. Then, for instance, the coefficient of the word *abbaa* in the series equals the upper right-hand corner entry in the product matrix $M_a M_b M_b M_a M_a$. Since $h(\lambda)$ must be the identity matrix and, consequently, we would necessarily have $(r, \lambda) = 0$ for $n > 1$, we allow a separate definition of the integer (r, λ). Instead of taking the upper right-hand corner entry, we could fix some other projection on the matrices. For such details the reader is referred to [Sa So]. This reference also contains proofs of the facts presented in the following lemma, as well as explanations concerning the term "Z-rational". (Indeed, our notion is customarily called Z-recognizable but the two notions define the same class of series.)

Lemma 3.11. The sum, difference, product and Hadamard product of two Z-rational formal power series is Z-rational and effectively constructable from the given series. The characteristic series of a regular language is Z-rational.

Example 3.6. Assume that $V = \{x\}$ and define

$$M_x = \begin{pmatrix} 0 & 1 \\ 1 & 1 \end{pmatrix}.$$

Consider the Z-rational formal power series

$$r = \sum_{i=0}^{\infty} p(M_x^i) x^i,$$

where p is the projection mapping a square matrix into its upper right-hand corner entry. (We do not define the entry (r, λ) separately and, consequently, $(r, \lambda) = 0$.) Denote the coefficients by a_i:

$$r = \sum_{i=0}^{\infty} a_i x^i.$$

(In the case of one-letter alphabets the coefficients naturally form a sequence, a *Z-rational sequence*.) It is easy to see that the coefficients form the *Fibonacci sequence*, $a_i = F_i$, where F_i is defined by

$$F_0 = 0, \quad F_1 = 1, \quad F_i = F_{i-2} + F_{i-1}, \quad \text{for } i \geq 2.$$

Indeed, $a_0 = 0$, $a_1 = 1$ and

$$M^i = \begin{pmatrix} F_{i-1} & F_i \\ F_i & F_{i+1} \end{pmatrix}, \quad \text{for all } i \geq 2.$$

This matrix equation can be verified inductively by the identity

$$M^{i+1} = \begin{pmatrix} 0 & 1 \\ 1 & 1 \end{pmatrix} M^i = \begin{pmatrix} 0 & 1 \\ 1 & 1 \end{pmatrix} \begin{pmatrix} F_{i-1} & F_i \\ F_i & F_{i+1} \end{pmatrix}$$

$$= \begin{pmatrix} F_i & F_{i+1} \\ F_{i-1} + F_i & F_i + F_{i+1} \end{pmatrix} = \begin{pmatrix} F_i & F_{i+1} \\ F_{i+1} & F_{i+2} \end{pmatrix}.$$

These equations also show that we could obtain the Fibonacci sequence from the powers of M by many other projections p as well. For instance, we could define p to be any of the four entries of the matrix, or either one of the two row sums or the two column sums. In each case the Fibonacci sequence results, possibly with a slight change in the initialization. $\square$

We will now establish several undecidability results concerning Z-rational formal power series by a reduction to Hilbert's Tenth Problem. The next lemma is the basic tool in the reduction.

Lemma 3.12. For any polynomial $P(x_1, \ldots, x_k)$ with $k \geq 1$ variables and (possibly negative) integer coefficients, a Z-rational formal power series r_P over the alphabet $V = \{a, b\}$ may be effectively constructed such that

$$(*) \qquad (r_P, a^{x_1} b a^{x_2} b \ldots a^{x_{k-1}} b a^{x_k}) = P(x_1, \ldots, x_k),$$

for all $x_1 \geq 0, \ldots, x_k \geq 0$. Moreover, $(r_P, w) = 0$ if w is not of the form mentioned in $(*)$.

Proof. The series $r(a) = \sum_{i=0}^{\infty} i a^i$ is Z-rational, since it can be defined by the matrix

$$M_a = \begin{pmatrix} 1 & 1 \\ 0 & 1 \end{pmatrix} \text{ satisfying } M_a^i = \begin{pmatrix} 1 & i \\ 0 & 1 \end{pmatrix}, \quad \text{for all } i \geq 0.$$

For each j, $1 \leq j \leq k$, consider the series r_j over the alphabet $\{a, b\}$ defined by

$$r_j = \text{char}((a^*b)^{j-1}) r(a) \, \text{char}((ba^*)^{k-j}).$$

Being a product of Z-rational series, r_j is Z-rational by Lemma 3.11. Moreover,

$$(r_j,\ a^{x_1}b\ldots ba^{x_k}) = x_j\,,$$

for all $x_1 \geq 0, \ldots, x_k \geq 0$. Indeed, only the exponent x_j, coming from $r(a)$, contributes a factor $\neq 1$ to the coefficient of r_j. The support of r_j consists of the words mentioned.

The series r_P is now constructed by forming sums, differences and Hadamard products of the series r_j, according to the formation rules of the polynomial P. For instance, if $P(x_1x_2) = 2x_1x_2^2 - x_1^2$, we obtain

$$r_P = r_1 \odot r_2 \odot r_2 + r_1 \odot r_2 \odot r_2 - r_1 \odot r_1\,.$$

□

Theorem 3.13. Each of the following problems (i)–(iv) is undecidable for a given Z-rational series r over the alphabet $V = \{a, b\}$:

(i) Is at least one coefficient of r equal to 0 ?
(ii) Does r have infinitely many coefficients equal to 0 ?
(iii) Does r have at least one positive coefficient ?
(iv) Does r have infinitely many positive coefficients ?

Proof. We show that if we could solve any one of problems (i)–(iv), we could also decide whether or not the equation

$$P(x_1, \ldots, x_k) = 0\,,$$

where P is a given polynomial with integer coefficients, has a solution in nonnegative integers.

Thus, let P be given. We construct the Z-rational series r_P according to Lemma 3.12. We will show how we can use an algorithm to solve some of problems (i)–(iv) for deciding whether or not the equation $P = 0$ has a solution in nonnegative integers.

Assume first that we can solve (i). Then we can decide whether or not the series

$$r_1 = r_P \odot r_P + \mathrm{char}(\sim((a^*b)^{k-1}a^*))\,,$$

which is Z-rational by Lemmas 3.11 and 3.12, possesses at least one coefficient equal to 0. But this happens exactly in the case where $P = 0$ has a solution in nonnegative integers. Indeed, in r_1 all coefficients belonging to words not of the form (*) of Lemma 3.12 are equal to 1. Other coefficients, that is, coefficients of the form

$$(r_1,\ a^{x_1}b\ldots ba^{x_k})$$

are either positive or 0, where the latter possibility occurs for k-tuples $(x_1, \ldots, x_k)$ such that $P(x_1, \ldots, x_k) = 0$. This argument also shows that the Z-rational series

$$r_3 = \mathrm{char}(V^*) - r_1$$

has a positive coefficient iff the equation $P(x_1, \ldots, x_k) = 0$ has a solution in nonnegative integers. Thus, if we could solve either (i) or (iii), we could also solve Hilbert's Tenth Problem. (Observe that the Hadamard product in the definition of r_1 is only needed for (iii).)

The reduction in the case of (ii) and (iv) is a slight modification of the reduction for (i) and (iii). We now define

$$r_2 = (r_P \text{ char}(b^*)) \odot (r_P \text{ char}(b^*)) + \text{char}(\sim ((a^*b)^{k-1}a^*b^*)),$$

$$r_4 = \text{char}(V^*) - r_2.$$

Then r_2 has infinitely many coefficients equal to 0 iff r_4 has infinitely many positive coefficients iff our equation $P = 0$ has at least one solution in nonnegative integers. □

Matters discussed in Theorem 3.13 become much more complicated if Z-rational sequences are considered, that is, the alphabet V has only one letter. Then problem (ii) is known to be decidable (but the proof is difficult), whereas the decidability status of (i), (iii) and (iv) is open. It is generally conjectured that the problems are decidable. As seen in our next theorem, the proof of this conjecture for problem (iii) is at least as difficult as the proof for problem (i). We will return to these matters in Section 4.1.

Theorem 3.14. Let r be an arbitrarily given Z-rational series over the alphabet $V = \{a\}$. Then any algorithm for solving problem (iii), as presented in the statement of Theorem 3.13, also yields an algorithm for solving problem (i).

Proof. Let r be an arbitrary Z-rational series over the alphabet $V = \{a\}$. By Lemma 3.11, the series $-(r \odot r) + \text{char}(V^*)$ is Z-rational as well. Since all coefficients in $-(r \odot r)$ are nonpositive, the whole series has a positive coefficient exactly in the case where r has a coefficient equal to 0. Hence, to decide the validity of the latter condition, it suffices to apply our algorithm for (iii) to the series $-(r \odot r) + \text{char}(V^*)$.
□

Matters discussed in Theorems 3.13 and 3.14 have a wide range of applications to the theory of L systems, models for biological development, [Ro Sa]. (Developmental models provided by L systems have turned out to be very suitable for computer graphics, see [Pr Li].) Some of the interconnections between L systems and matters discussed in Theorems 3.13 and 3.14 will be dealt with in Section 4.1.

We conclude this chapter with a somewhat different application of Hilbert's Tenth Problem to developmental models. We first consider an introductory example.

Example 3.7. Consider filamentous (string-like) organisms, consisting of four types of cells, referred to as "blue", "red", "white" and "yellow" cells, abbreviated b, r, w, y. Some cell divisions occur. During the day each blue cell divides into blue and red, that is, becomes the string br. Similarly, w becomes wy, whereas r

and y remain unchanged during the day. We represent the daytime developments symbolically as follows.

$$D: \ b \to br \ , \ r \to r \ , \ w \to wy \ , \ y \to y \ .$$

During the night the development is somewhat different, and is represented symbolically as follows:

$$N: \ b \to bw \ , \ r \to ry \ , \ w \to w \ , \ y \to y \ .$$

Thus, the yellow cells never change.

The development of the filament $brwy$ for two days and nights can be written as

$$brwy \Rightarrow_D brrwyy \Rightarrow_N bwryrywyy$$

$$\Rightarrow_D brwyryrywyyy \Rightarrow_N bwrywyryyryywyyy \ .$$

If we interpret D as a period under illumination, and N as a period without illumination, it is not necessary to alternate D and N. We obtain, for instance, the following developments:

$$brwy \Rightarrow_N bwrywy \Rightarrow_N bwwryywy$$

$$\Rightarrow_D brwywyryywyy \Rightarrow_D brrwyywyyryywyyy \ ;$$

$$brwy \Rightarrow_N bwrywy \Rightarrow_D brwyrywyy$$

$$\Rightarrow_D brrwyyrywyyy \Rightarrow_N bwryrywyyryywyyy \ .$$

In each of the three developmental sequences, two D periods and two N periods occur. We observe that, although there are differences in the order of the cells, the final filament in each of the three sequences contains one b-cell, nine y-cells, and three r-cells and three w-cells. Thus, in each of the three cases we obtain the Parikh vector $(1, 3, 3, 9)$.

This also holds true in general: the Parikh vector of the final filament depends only on the number of D- and N-periods, not on their mutual order. Instead of the initial filament $brwy$, we may consider any other initial filament, and the commutativity between D and N still holds. This follows because the growth matrices M_D and M_N, based on the cell division rules, commute.

Indeed, consider the matrices

$$M_D = \begin{pmatrix} 1 & 1 & 0 & 0 \\ 0 & 1 & 0 & 0 \\ 0 & 0 & 1 & 1 \\ 0 & 0 & 0 & 1 \end{pmatrix} , \ M_N = \begin{pmatrix} 1 & 0 & 1 & 0 \\ 0 & 1 & 0 & 1 \\ 0 & 0 & 1 & 0 \\ 0 & 0 & 0 & 1 \end{pmatrix} .$$

Let h be the morphism of $\{D, N\}^*$ into the multiplicative monoid of four-dimensional square matrices with nonnegative integer entries, defined by

$$h(D) = M_D \;,\; h(N) = M_N \;.$$

Then it is easy to show inductively that, for any $x \in \{D, N\}^*$, the Parikh vector of the filament produced after a sequence of Ds and Ns specified by x equals $\psi_0 h(x)$, where ψ_0 is the Parikh vector of the initial filament. (Above the initial vector is $brwy$ with $\psi_0 = (1,1,1,1)$.)

The matrices M_D and M_N commute:

$$M_D M_N = M_N M_D = \begin{pmatrix} 1 & 1 & 1 & 1 \\ 0 & 1 & 0 & 1 \\ 0 & 0 & 1 & 1 \\ 0 & 0 & 0 & 1 \end{pmatrix} \;.$$

From this result we obtain the claim made above: for any ψ_o and any x and y with $\psi(x) = \psi(y)$,

$$\psi_o h(x) = \psi_o h(y) \;.$$

Of course, commutativity is a rather exceptional property of matrices, caused here by our particular cell division rules for day and night. It is to be emphasized that the matrices M_D and M_N only keep track of the number of occurrences of each of the cell types; all information concerning the order of the cells is lost. □

We now define the notion of a *DTOL function* formally. Let n be a positive integer, π an n-element row vector with nonnegative entries, and η an n-element column vector with all entries equal to 1. Let h be a morphism of V^*, where V is an alphabet, into the multiplicative monoid of n-dimensional square matrices with nonnegative integer entries. Then the mapping f of V^* into the set N of nonnegative integers, defined by

$$f(x) = \pi h(x) \eta \;,$$

is termed a *DTOL function*. A *DTOL* function f is termed *commutative* iff

$$f(x) = f(y) \text{ whenever } \psi(x) = \psi(y) \;.$$

For instance, the matrices M_D and M_N, as well as the initial vector $(1,1,1,1)$ considered in Example 3.7, define a commutative *DTOL* function f. It is easy to see that

$$f(x) = (|x|_D + 2)(|x|_N + 2) \;,$$

where $|x|_D$ is the number of Ds in x, and $|x|_N$ is the number of Ns in x.

Because of the final vector η, *DTOL* functions express only the length of a word (final filament), not its Parikh vector. In Example 3.7, the matrices M_D and M_N

commute. In general, the pairwise commutativity of the matrices $h(a)$, where a ranges over V, is sufficient but by no means necessary for the commutativity of f. Even if the matrices do not commute, commutativity of f may still result because of the projection due to π and η. One can show, by an argument based on linear algebra, that commutativity of f is a decidable property.

One may also consider formal power series

$$\Sigma(r,w)w\ ,\ (r,w) = f(w)\ ,$$

where f is a $DTOL$ function. Such series, especially if f is commutative, form a very special subclass of Z-rational series. Still, as will be seen in our next lemma, the expressive power of this restricted subclass allows reductions to Hilbert's Tenth Problem.

Lemma 3.15. For every polynomial $P(y_1, \ldots, y_k)$ with integer coefficients, one can effectively find a constant m_0 such that, for any integer $m \geq m_0$, the function $f: \{a_1, \ldots, a_n\}^* \to N$ defined by

$$f(x) = (|x| + k + 1)^m + P(|x|_{a_1}, \ldots, |x|_{a_k})$$

is a commutative $DTOL$ function.

The proof of Lemma 3.15 is omitted. The proof uses techniques from both Z-rational series and $DTOL$ systems. The "dominant term" $(|x|+k+1)^m$ guarantees that f does not assume negative values. The dominant term also takes care of the negative contributions to the length caused by negative coefficients in P. In our cell division model negative growth occurs if a cell dies, that is, a letter becomes the empty word. Negative contributions of P are taken care of by including a sufficient amount of dying letters in that part corresponding to the dominant term. If all coefficients of P are nonnegative, then no dominant term is needed; the function

$$f(x) = P(|x|_{a_1} + 1, \ldots, |x|_{a_k} + 1)$$

is a commutative $DTOL$ function.

Observe that the growth of a commutative $DTOL$ function f may be exponential. Thus, there is no constant t such that $f(x) \leq |x|^t$ holds for all x. However, the undecidability results of our final theorem in this chapter hold true even if only polynomially bounded functions are considered. The theorem only presents some sample results, among many similar ones, where a reduction to Hilbert's Tenth Problem is used.

Theorem 3.16. The following problems are undecidable for two given polynomially bounded commutative $DTOL$ functions f and g:

(i) Is there an x such that $f(x) = g(x)$?
(ii) Does $f(x) \geq g(x)$ hold for all words x ?

Proof. We apply Lemma 3.15 in the reduction. Clearly, the function f of the lemma is polynomially bounded, and the function $(|x|+k+1)^m$ is a (polynomially bounded commutative) $DTOL$ function. (i) is established by taking the function f of the lemma and
$$g(x) = (|x|+k+1)^m \ .$$
(Here m is fixed for both g and f; for instance, $m = m_0$.) To establish (ii), we use the functions
$$\begin{aligned} f(x) &= (|x|+k+1)^m + (P(|x|_{a_1},\ldots,|x|_{a_k}))^2 \ , \\ g(x) &= (|x|+k+1)^m + 1 \ . \end{aligned}$$
□

With the topic of the next section already in mind, we finally mention that the problem "Is there a word x and a letter b such that $f(x) = f(xb)$" is undecidable for polynomially bounded $DTOL$ functions f but the decidability is open for polynomially bounded commutative $DTOL$ functions f.

Interlude 2

Bolgani. By now you have also read Chapters 2 and 3. Many of the issues dealt with in Chapters 1–5 are of fundamental importance in mathematics and computer science. Moreover, we are dealing here with problems that are also of general philosophical interest to the layman. What does it mean to say that results in mathematics are "exact" or "certain", and to what extent are such claims justified? Is mathematics discovery or creation? When somebody publishes a theorem as a new result, has he or she only uncovered something that existed before? Is there a world (that could be called a Platonic world) where all theorems are there for us to discover? Such issues come to mind when we talk about undecidability, Gödel sentences, etc. Questions about polynomials with integer coefficients, dealt with in Chapter 3, belong to basic arithmetic. It is reasonable to expect that they would have definite, absolute and sure answers. That this is not the case might be a mental earthquake for many. Not much of mathematics is of such general interest.

Tarzan. Such all-embracing fundamental issues certainly have a long history behind them. Some references and hints to the literature were given in the text.

Bolgani. It was all I wanted to do. Tracing back the history of everything that is nowadays commonly regarded as folklore would constitute a treatise on its own. Extensive bibliographical comments would have decreased the general readability. Besides, much of the material in Chapter 2 has not appeared before. At least not in the present formulation, and the same remark applies to some other parts of the book as well. At least [Thue], [Post 1], [Post 3], [Göd 1], [Tur] and [Chu] should be mentioned as fundamental papers with regard to the contents of this book. [Jo Ma] was very useful for Chapter 3, and so were [Cha 1] and [Cal] for Chapter 5 — Cris Calude kindly provided an early version of the latter for my disposal. Besides, I would like to mention [Rog], [Min] and [Sal 1] as extensive early expositions on recursive functions, computability and languages, respectively.

It would be a futile attempt to try to acknowledge all those people who have contributed to the book. We have discussed the topics with many people. I only mention here that Lila Kari and Alexandru Mateescu have made useful comments about some parts of the text.

Tarzan. And the role of Elisa Mikkola was very important. She transformed versions of text resembling apes' handwriting into LaTeX format. But I would like to return now to the general discussion about formal systems, mathematics and undecidability.

Kurt. According to the formalist view one adjoins to the meaningful propositions of mathematics transfinite (pseudo-)assertions, which in themselves have no meaning, but serve only to round out the system, just as in geometry one rounds out a system by the introduction of points at infinity. This view presupposes that, if one adjoins to the system S of meaningful propositions the system T of transfinite propositions and axioms and then proves a theorem of S by making a detour through theorems of T, this theorem is also contentually correct, hence that through the adjunction of transfinite axioms no contentually false theorems become provable. This requirement is customarily replaced by that of consistency. Now I would like to point out that one cannot, without further ado, regard these two demands as equivalent. For, if in a consistent formal system A (say that of classical mathematics) a meaningful proposition p is provable with the help of transfinite axioms, there follows from the consistency of A only that not-p is not formally provable *within* the system A. None the less it remains conceivable that one could ascertain not-p through some sort of contentual considerations that are not formally representable in A. In that case, despite the consistency of A, there would be provable in A a proposition whose falsity one could ascertain through finitary considerations. To be sure, as soon as one interprets the notion "meaningful proposition" sufficiently narrowly (for example, as restricted to finitary numerical equations), something of that kind cannot happen. However, it is quite possible, for example, that one could prove a statement of the form $(\exists x)F(x)$, where F is a finitary property of natural numbers (the negation of Goldbach's conjecture, for example, has this form), by the transfinite means of classical mathematics. On the other hand one could ascertain by means of contentual considerations that all numbers have the property not-F; indeed, and here is precisely my point, this would still be possible even if one had demonstrated the consistency of the formal system of classical mathematics. For of no formal system can one affirm with certainty that all contentual considerations are representable within it.

Bolgani. Hilbert's program might have seemed to be doomed by Gödel's second incompleteness theorem. However, Hilbert's program was continued by using new principles proposed as being finitary. The computational or algorithmic aspect was not yet taken into account.

Kurt. Let us consider the concept of demonstrability. It is well known that, in whichever way you make it precise by means of a formalism, the contemplation of this very formalism gives rise to new axioms which are exactly as evident and justified as those with which you started, and that this process of extension can be iterated into the transfinite. So there cannot exist any formalism which would embrace all these steps; but this does not exclude the fact that all these steps (or at least all of them which provide something new for the domain of propositions in which you are interested) could be described and collected together in some non-constructive way.

David. If the arbitrarily given axioms do not contradict each other through their consequences, then they are true, then the objects defined through the axioms exist. That, for me, is the criterion for truth and existence.

The paradoxes had the intolerable effect that one started to think that even in mathematics, which is considered to be a model of certainty and truth, something improper comes out of concepts and arguments learned, taught and applied by everybody. Where can one find certainty and truth if not in mathematical thinking?

However, there is a completely satisfactory way out of this situation. We always have to look for, develop, support and make as applicable as possible new concepts and proof methods. Nobody is going to drive us out from the paradise created for us by Cantor.

Bolgani. Very optimistic views. However, if one keeps in mind the algorithmic point of view, the proof methods cannot be extended indefinitely. Let us go back to Church's Thesis, formulated originally in [Chu]. It was not widely accepted at the beginning, and was also doubted by Gödel who, however, later spoke very differently than he had before. In [Tur] Turing defended the Thesis by three kinds of arguments: a direct appeal to intuition, a proof of the equivalence of two definitions of an algorithm (in case the new definition has a greater intuitive appeal), and by examples of large classes of algorithms covered by the formal definition. Now when the Thesis is generally accepted, we may consider it a law of nature that, for instance, proofs in a formal system can be checked by a Turing machine. Indeed, we can identify a formal system with a Turing machine.

Kurt. In consequence of later advances, in particular of the fact that due to A. M. Turing's work a precise and unquestionably adequate definition of the general notion of formal system can now be given, completely general versions of incompleteness theorems are now possible. Thus, it can be proved rigorously that in *every* consistent formal system that contains a certain amount of finitary number theory there exist undecidable arithmetic propositions and that, moreover, the consistency of any such system cannot be proved in the system. In my opinion the term "formal system" or "formalism" should never be used for anything but this notion. Earlier I suggested certain transfinite generalizations of formalisms; but these are something radically different from formal systems in the proper sense of the term, whose characteristic property is that reasoning in them can, in principle, be completely replaced by mechanical devices.

Emil. The unsolvability of the finiteness problem for all normal systems, and the essential incompleteness of all symbolic logics are evidence of limitations in man's mathematical powers, creative though these may be. They suggest that in the realms of proof, as in the realms of process, a problem may be posed whose difficulties we can never overcome; that is, that we may be able to find a definite proposition which can never be proved or disproved. The theme will intrude every so often in our more immediate task of obtaining an analysis of finite processes.

Tarzan. I have often heard people mentioning the undecidability results as evidences that man is something more than a machine.

Emil. The conclusion that man is not a machine is invalid. All we can say is that man cannot construct a machine which can do all the thinking he can do. To illustrate this point we may note that a kind of machine-man could be constructed who would prove a similar theorem for his mental acts.

The creative germ seems to be incapable of being purely presented but can be stated as consisting of constructing ever higher types. These are as transfinite ordinals and the creative process consists of continually transcending them by seeing previously unseen laws which give a sequence of such numbers. Now it seems that this complete seeing is a complicated process and mostly subconscious. But it is not given till it becomes completely conscious. But then it ought to be purely mechanically constructable.

Tarzan. This suggests that the Platonic world from which we uncover things is in our subconscious minds. But is it the same world in your mind as in mine? Is the resulting activity uniform?

Emil. We have to do with a certain activity of the human mind as situated in the universe. As activity, this logico-mathematical process has certain temporal properties; as situated or performed in the universe it has certain spatial properties.

Now the objects of this activity may be anything in the universe. The method seems to be essentially that of symbolization. It may be noted that language, the essential means of human communication, is just symbolization. In so far as our analysis of this activity is concerned, its most important feature is its ability to be self-conscious.

A summary of the method is as follows. Try to give a complete description of what goes on. In this description we symbolize everything. This symbolizes away most things. But a few things cannot be symbolized away, because in the transformation produced by the symbolization they reappear of necessity. These things are meaning, symbolization, symbol manipulation, iteration, sense perception or direct verification, and perhaps a few other things. These will then constitute the elements out of which the description is built, in addition to mere symbols.

Chapter 4

Classes of Problems and Proofs

4.1 Problems with many faces

The purpose of this chapter is to ramify and extend the discusions in the previous chapters, rather than to introduce a new reference point for undecidability proofs.

Different-looking problems can be equivalent from the point of view of decidability: one of them is decidable exactly in the case where the other is decidable. This will be the subject matter of the present section. Many times it has been the case that the equivalence (in the sense mentioned) of several problems has been known long before the decidability status has been established — if it has been established at all. Thus, one has to make a choice of which among the different equivalent variations is the most suitable for a closer investigation as regards decidability. The heading of this section could also be "Decision problems and problem decisions".

The state of affairs that the equivalence of several variations is known, although the decidability status is open, is very natural. Equivalence means simply the knowledge of a mapping, reduction, between the sets involved. This knowledge as such does not yet provide any information about the eventual recursiveness of the sets. The decidability of a problem, such as the Post Correspondence Problem, means that both the "yes-instances" and the "no-instances" constitute recursively enumerable sets. As regards the Post Correspondence Problem, one can recursively enumerate all yes-instances by straightforward dovetailing. There is no such straightforward enumeration for the no-instances. Indeed, we know it is impossible by Theorem 2.1.

Before actually beginning our discussion about the many faces of the same problem, we will say a few words concerning the contents of this chapter in general. Section 4.2 deals with the borderline between decidability and undecidability. A parameter might be involved in the formulation of a problem. If the parameter is numerical, there is sometimes a cut-point value of the parameter, where the problem changes from decidable to undecidable. Knowledge of such a cut-point value often constitutes only a very rough estimate about the borderline, and other measures are called for to obtain a sharper view. The final section of this chapter discusses different notions of mathematical proofs, in particular zero-knowledge proofs that open rather surprising vistas for communication.

We now begin the presentation of different-looking variations of the same problem. This is to be understood in the sense explained above: the variations are equivalent as far as decidability is concerned. No general theory is aimed at — our purpose is to present some typical illustrations. Attention will be focused on the aforementioned cases, where the historical perspective is visible: equivalence was known before decidability status. All three possibilities concerning decidability status will be met:

(i) The problem turned out to be undecidable.
(ii) The problem turned out to be decidable.
(iii) The decidability status is still open.

A very typical example of case (i) was met in Chapter 3. We are looking for an algorithm to decide whether or not a polynomial equation with integer coefficients

(*) $$P(x_1, \ldots, x_n) = 0$$

has a solution in integers. From the decidability point of view, it is equivalent to asking whether or not (*) has a solution in nonnegative integers. This follows because (*) has a solution in nonnegative integers iff the equation

$$P(p_1^2 + q_1^2 + r_1^2 + s_1^2, \ldots, p_n^2 + q_n^2 + r_n^2 + s_n^2) = 0$$

has an integral solution. On the other hand, (*) has a solution in integers iff the equation

$$P(p_1 - q_1, \ldots, p_n - q_n) = 0$$

has a solution in nonnegative integers. This equivalence was known long before undecidability.

Our emphasis will be on language-theoretic problems. In the examples in the following simple theorem, one of the variants is actually a subproblem of the other — so the decidability of the latter (resp. the undecidability of the former) immediately implies the decidability of the former (resp. the undecidability of the latter).

Theorem 4.1. The emptiness problem for languages $L(G)$ generated by length-increasing grammars G is decidable iff the emptiness problem for languages $L(G)$ generated by arbitrary grammars G is decidable. If the equivalence problem for OL languages is decidable, then so is the equivalence problem for languages of sentential forms generated by context-free grammars.

Proof. Consider the first sentence. Since length-increasing grammars constitute a special case of arbitrary grammars, it suffices to show that an algorithm for deciding the emptiness of languages $L(G')$, where G' is length-increasing, can be transformed into an algorithm for deciding the emptiness of languages $L(G)$, where G is arbitrary. This can be done as follows.

We transform an arbitrary grammar G into a length-increasing grammar G'. In addition to the letters of G, G' contains two new terminal letters, say a and b, and two new nonterminals S' and X, the former being the initial letter of G'. Let $u \to w$ be a production of G. If $|u| \leq |w|$, then G' contains the same production $u \to w$. Otherwise, G' contains the production $u \to wX^{|u|-|w|}$. Moreover, G' contains the productions

$$S' \to bS \,,\ \alpha X \to X\alpha \,,\ bX \to ab \,,$$

where S is the initial letter of G, and α ranges over all letters of G.

Clearly, G' is length-increasing because of the "filling" effect of X. Moreover, X has no other function; it travels to the left end of the word, where it changes to a. The original derivations according to G can also be carried out in G', after the occurrences of X have been transferred to the beginning of the word. This means that $L(G')$ consists of words of the form $a^i bw$, where w is in $L(G)$. Moreover, for each word w in $L(G)$, there is in $L(G')$ a word of the form mentioned. Consequently, $L(G')$ is empty iff $L(G)$ is empty, and the algorithm concerning G' becomes applicable.

The second sentence was essentially proved in Theorem 2.12 (iv). However, we were dealing then with linear grammars only — so let us modify the argument for the present case. We transform each context-free grammar G into an OL system H by disregarding the difference between terminals and nonterminals and adding the production $\alpha \to \alpha$ for every letter α. Then $L(H)$ equals the language of sentential forms of G. This follows because the new productions are irrelevant in G, whereas they destroy the parallelism in H. In this way, any algorithm for deciding the equivalence of OL languages becomes applicable for deciding the equivalence of languages of sentential forms.

□

Could we establish the converse of the second sentence of Theorem 4.1? "If the equivalence problem for languages of sentential forms generated by context-free grammars is decidable, then so is the equivalence problem for OL languages." As such this is a true statement, since both problems are undecidable by Theorem 2.12. However, the situation is much trickier if we take the statement to mean (analogously as in the proof of Theorem 4.1) that any algorithm for the former problem can be transformed into an algorithm for the latter problem. We now illustrate this issue.

Let us look into this question in the simple case where the alphabet consists of one letter only. In this case we speak of UOL and USF languages. Thus, a USF language is generated using productions $a \to a^i$, $i \geq 0$, in a sequential fashion. Instead of only a, we allow the initial word to be of the form a^j. In this way, we have uniformity of initial words in parallel and sequential cases.

Example 4.1. Assume that a^2 is the initial word and we have the productions $a \to a^2$ and $a \to a^5$. If we view this setup as a system generating a USF language, then the language will be $\{a^i | i \geq 2\}$ and, in fact, the second production is superfluous. On the other hand, if we are dealing with a UOL system, it is easy to show inductively that the generated language is $\{a^i | i \geq 2,\ i \neq 5$ and 3 does not divide $i\}$.

□

Theorem 4.2. The equivalence problem for USF languages is decidable.

Proof. Assume first that $a \to \lambda$ is present in one of the given systems. If it is not present in the other system, the languages cannot be the same (except for the trivial case where both languages consist of the empty word alone). On the other hand, the presence of $a \to \lambda$ implies that the generated language is either a^* or of the form $\{a^i | i \leq k\}$. Which alternative holds, as well as the value of k, can be immediately determined. Hence, we may assume that in both systems the productions are of the form $a \to a^i$, $i \geq 2$. (The production $a \to a$ can be omitted in the USF case.) We may also assume that both systems have the same initial word a^j because, otherwise, we immediately conclude that the languages are not the same.

We now order the exponents of the right sides of the productions:

$$2 \leq i_1 < i_2 < \ldots < i_r \text{ and } 2 \leq j_1 < j_2 < \ldots < j_s,$$

where the two sequences come from the two systems. We call a number i_k *superfluous* if $i_k - 1$ is a linear combination of the numbers $i_1 - 1, \ldots, i_{k-1} - 1$. The productions, where the exponent is superfluous, can be removed without affecting the language. We assume that superfluous numbers have indeed been removed from our two sequences above. (For instance, i_3 is superfluous if $i_1 = 6$, $i_2 = 12$, $i_3 = 22$. Indeed, the effect of the production $a \to a^{22}$ is reached by applying twice the production $a \to a^6$ and once $a \to a^{12}$.)

It is now obvious by inductive argument that the languages are the same only if the systems are identical, that is, $r = s$ and $i_1 = j_1$, ..., $i_r = j_s$.

□

The above argument can also be expressed by saying that every USF language is generated by a system in a unique "normal form". The algorithm for deciding the equivalence consists of reducing both systems to the normal form and checking whether or not the normal forms coincide.

Let us now return to our question about the converse of the second sentence of Theorem 4.1. The equivalence problem is decidable for UOL languages as well. However, the algorithm is rather involved, and there is no way of transforming the simple algorithm of Theorem 4.2 into this involved one. Specifically, the transformation would have to contain elements intimately connected with the involved algorithm itself. The next example shows some of the difficulties. It should be remembered that we are now dealing with the one-letter case. Of course, the difficulties are greater in the case of a general alphabet. The family of USF languages is a small subfamily of UOL languages. The corresponding statement also holds for general alphabets. Thus, both sentences of Theorem 4.1 compare a problem with its subproblem but there is a remarkable difference as regards the transformation of the decision methods.

Example 4.2. This example resembles Example 4.1. Take the UOL system with axiom a and productions $a \to a^2$ and $a \to a^5$. The generated language will be
$$\{a^i | i \geq 1 \text{ and } 3 \text{ does not divide } i\}.$$
What about adding the production $a \to a^7$? Then we also obtain the word a^9 that we did not have before. Indeed, the new language will be
$$\{a^i | i \geq 1 \text{ and } i \neq 3, 6, 12\}.$$
The same language also results with the productions $a \to a^2$, $a \to a^5$ and $a \to a^9$. Thus, the two UOL systems with axiom a and production set
$$\{a \to a^2, a \to a^5, a \to a^7\} \text{ or } \{a \to a^2, a \to a^5, a \to a^9\}$$
are equivalent. However, the algorithm of Theorem 4.1 is not very helpful for showing the equivalence: it entirely misses the modular arithmetic needed. There are even much more complicated phenomena occurring in this modular arithmetic. □

Our next illustration is a very famous problem for which many different versions were known to be equivalent from the point of view of decidability, until the problem was finally shown to be decidable. The "many faces" of this problem that will be discussed below are: the *reachability problem for vector addition systems*, the *semi-Thue problem for Abelian semigroups*, the *emptiness problem for matrix languages* and the *reachability problem for Petri nets*.

By definition, an *n-dimensional vector addition system*, $n \geq 2$, is a finite set
$$VAS = \{v_1, \ldots, v_r\},$$
where each v_i is an ordered n-tuple of integers, referred to as a *vector*. Let u and u' be vectors with nonnegative components. Then u' is *reachable* from u according to VAS iff there is a finite sequence
$$u_0 = u, u_1, \ldots, u_m = u'$$
of vectors with *nonnegative components* such that for each i, $1 \leq i \leq m$, there is a j, $1 \leq j \leq r$, with the property
$$u_i = u_{i-1} + v_j.$$
(The addition of vectors is carried out componentwise.) Thus, v_j may have negative components but the components of each u_i must be nonnegative.

Example 4.3. Assume that $n = 2$ and define

$$VAS = \{(1, 0), (0, 1), (-100, -100)\} \, .$$

Then every u' is reachable from every u according to VAS. Define

$$VAS' = \{(1, -2), (-3, 1)\} \, .$$

Then $(0, 0)$ is not reachable from $(5, 0)$, although it can be reached via the illegal sequence

$$(5, 0), (2, 1), (-1, 2), (0, 0) \, .$$

□

The reachability problem for vector addition systems consists of finding an algorithm that will decide for an arbitrary triple (VAS, u, u') whether or not u' is reachable from u according to VAS. The key issue is that no negative components are allowed in the intermediate steps. This prevents the use of standard linear algebra.

We now describe another variant of the problem. Consider an alphabet Σ and a finite set of pairs of words over Σ, written as productions:

(*) $\qquad\qquad\qquad u_1 \to v_1, \ldots, u_k \to v_k \, .$

We can now speak of the yield relation and derivations as usual. However, the additional feature is that letters *commute*, that is, the order of letters is irrelevant. We can define this formally by agreeing that $ab \to ba$ is among the productions, for each letter a and b.

We want to construct an algorithm for deciding for a given triple $((*), w, w')$, where w and w' are words over Σ, whether or not $w \Rightarrow^* w'$ holds in the rewriting system specified by (*). This problem is referred to as the *semi-Thue problem for Abelian semigroups*. (According to conventional terminology, the semi-Thue problem refers to a modification of the word problem, where unidirectional rewriting rules $u \to v$ are applied instead of equations $u = v$.)

Theorem 4.3. The reachability problem for vector addition systems and the semi-Thue problem for Abelian semigroups are equivalent with respect to decidability.

Proof. We consider Parikh vectors

$$\psi(w) = (i_1, \ldots, i_n)$$

associated with words over an alphabet $\Sigma = \{a_1, \ldots, a_n\}$. Conversely, if $v = (i_1, \ldots, i_n)$ is a vector with nonnegative components, we define the associated word by

$$W(v) = a_1^{i_1} \ldots a_n^{i_n} \, .$$

Assume first that an algorithm for the semi-Thue problem for Abelian semigroups is known. Then the reachability problem can be solved as follows. Given a triple (VAS, u, u'), we consider the triple $((*), W(u), W(u'))$, where the productions $(*)$ are based on VAS. More specifically, a vector v in VAS gives rise to a production $W(v') \to W(v'')$, where v' (resp. v'') is obtained from v by replacing positive components by 0 and negative components x by $-x$ (resp. by replacing negative components by 0). Thus, v' states what should be removed, and v'' what should be added. The overall effect is precisely that of v. Hence, u is reachable from u' exactly in case

$$W(u) \Rightarrow^* W(u').$$

Commutativity is of course essential in the argument.

Conversely, assume that an algorithm for the reachability problem is known. Consider the semi-Thue problem. Given $(*)$ and two words w and w', how do we decide whether or not $w \Rightarrow^* w'$ holds? As before, assume that $\Sigma = \{a_1, \ldots, a_n\}$. We define an $(n+3)$-dimensional vector addition system VAS. For each production $u_i \to v_i$, $1 \leq i \leq k$, VAS contains two vectors of dimension $n + 3$

$$(-\psi(u_i), -1, i, 2k + 1 - i) \text{ and } (\psi(v_i), 1, -i, i - 2k - 1).$$

(The notation $-\psi(u_i)$ should be obvious: the first n components are of the form $-x$, where x is the corresponding component of the Parikh vector.)

It is now easy to see that $w \Rightarrow^* w'$ holds iff the vector $(\psi(w'), 1, 0, 0)$ is reachable from the vector $(\psi(w), 1, 0, 0)$ according to VAS. In fact, the three last components in a permissible computation are always

$$(1, 0, 0) \text{ or } (0, i, 2k + 1 - i),$$

for some $i = 1, \ldots, k$. If the first alternative holds, we must add the first of the vectors corresponding to $u_i \to v_i$, for some i, in order to avoid negative components. After this the second alternative holds. Now we have to add the second of the vectors corresponding to $u_i \to v_i$. If we switch to a different value of i, one of the last two components will become negative. After adding the second vector, we are back in the first alternative.

□

A *matrix grammar* consists of finitely many finite sequences

$$[A_1 \to \alpha_1, \ldots, A_r \to \alpha_r]$$

(*matrices*) of context-free productions. An application of such a matrix to a word consists of replacing an occurrence of A_1 by α_1, and so forth until, finally, an occurrence of A_r is replaced by α_r. An entire matrix always has to be applied. If some production in a matrix is not applicable in its turn, the derivation halts without producing a terminal word.

Thus, a certain number of nonterminals has to be present before a specific matrix M can be applied. Assume that $\{A_1, \ldots, A_n\}$ is the alphabet of nonterminals. By scanning through M, we find an n-dimensional vector v_M such that M is applicable to a word α iff $\psi(\alpha) \geq v_M$, when only occurrences of nonterminals are considered in Parikh vectors. We omit the formal definition of v_M, referred to as the *precondition* for M. The definition should be clear from the following example.

Example 4.4. Assume that $n = 3$ and consider the matrix

$$M = [A_1 \to \lambda, A_1 \to \lambda, A_2 \to A_1 A_2 A_1, A_3 \to A_3].$$

Now $v_M = (2, 1, 1)$. Observe that M does not change the number of occurrences of any nonterminal. Next consider

$$M' = [A_1 \to A_2 A_3, A_2 \to \lambda, A_3 \to \lambda].$$

In this case $v_{M'} = (1, 0, 0)$: the necessary occurrences of A_2 and A_3 are created by the first production.

□

Theorem 4.4. The emptiness problem for languages generated by matrix grammars and the reachability problem for vector addition systems are equivalent with respect to decidability.

Proof. Terminal letters are irrelevant when one only wants to decide on the emptiness of the language. It suffices to consider Parikh vectors associated with nonterminals. If A_1 is the initial letter, then the generated language is nonempty iff the vector $(0, 0, \ldots, 0)$ can be obtained in a derivation from the vector $(1, 0, \ldots, 0)$. The order of nonterminals is irrelevant. The proof now resembles that of Theorem 4.3.

Assume that an algorithm for deciding the emptiness problem is known. Consider an arbitrary n-dimensional VAS. We construct a matrix grammar with n nonterminals. For each vector v in VAS, we construct a matrix M such that:

(i) M affects the change in the number of occurrences of nonterminals specified by v;
(ii) in v_M, the precondition for M, the negative components x of v appear in the form $-x$, whereas the positive components of v have been replaced by 0.

That this is always possible should be obvious from the following example. Assume that $n = 4$ and $v = (-1, 3, 2, -4)$. Then we choose

$$M = [A_1 \to \lambda, A_4 \to \lambda, A_4 \to \lambda, A_4 \to \lambda, A_4 \to A_2^3 A_3^2].$$

The idea is that all deletions are carried out first, and new letters are inserted with the last deletion. This technique does not work if all the components of v are nonnegative. Such vectors are handled by a special nonterminal B, for which the grammar contains the matrices $[A_1' \to BA_1]$, $[B \to \lambda]$ and $[B \to BW(v)]$, for each such v. Here A_1' is the initial letter for the matrix grammar and $W(v)$ is defined as in the proof of Theorem 4.3. The two additional nonterminals A'_1 and B are only introduced if there are such vectors v in VAS — otherwise the number of nonterminals equals n.

Clearly, the language generated by the matrix grammar is nonempty iff the vector $(0, 0, \ldots, 0)$ is reachable from the vector $(1, 0, \ldots, 0)$ in VAS. It is an easy exercise to reduce the arbitrary reachability problem to the reachability problem between these two specific vectors. We omit the details.

Conversely, assume that an algorithm for deciding the reachability problem is known. Consider a matrix grammar with n nonterminals and k matrices. We construct an $(n+3)$-dimensional VAS in almost the same way as in the proof of Theorem 4.3. For each of the k matrices M_i, $1 \leq i \leq k$, we consider the precondition v_{M_i} as defined above. Moreover, we let u_{M_i} be the n-dimensional vector whose jth component, $1 \leq j \leq n$, equals the change (positive or negative integer or 0) in the number of occurrences of A_j caused by one application of M_i. The idea is first to subtract v_{M_i} (in order to guarantee that the precondition for M_i is met), and then add the contribution brought forward by an application of M_i. After subtracting v_{M_i}, we may have subtracted more than is indicated by the negative components of u_{M_i}. This has to be corrected by adding $u_{M_i} + v_{M_i}$ rather than simply the positive part of u_{M_i}. For the matrices in Example 4.4, we have

$$u_M = (0, 0, 0) \text{ and } u_{M'} = (-1, 0, 0).$$

We are now ready to define the vector addition system. For each of the k matrices M_i, VAS contains the two vectors

$$(-v_{M_i}, -1, i, 2k + 1 - i) \text{ and } (u_{M_i} + v_{M_i}, 1, -i, i - 2k - 1).$$

It is now seen, in the same way as in Theorem 4.3, that the language generated by our matrix grammar is nonempty exactly in case the vector

$$(0, 0, \ldots, 0, 1, 0, 0)$$

is reachable from the vector

$$(1, 0, \ldots, 0, 1, 0, 0).$$

□

Another variant of our problem, equivalent as regards decidability, is the *reachability problem for Petri nets*. Petri nets are models of concurrency and parallelism which are very useful, for instance if one wants to model systems where many processors operate independently and in parallel and where some partial computations depend (perhaps in a complicated way) on the outcome of some other computations. Since there is a huge literature dealing with Petri nets and since the reduction does not contain any proof-theoretical novelties in comparison with our discussion above, we omit all details concerning the equivalence of the reachability problems for Petri nets and vector addition systems. The equivalence between all four problems (the two reachability problems, the semi-Thue problem for Abelian semigroups and the emptiness problem for matrix languages) was known long before the problems were shown to be decidable. Some of the proofs showing decidability turned out later to be wrong.

A notion central to the remainder of this section will be that of a *DOL system*. By definition, a *DOL* system is a triple

$$G = (\Sigma, h, w_0),$$

where Σ is an alphabet, $h : \Sigma^* \to \Sigma^*$ is a morphism and w_0 is a word over Σ, called the *axiom*. A *DOL* system generates a *sequence*, $E(G)$, obtained by iterating the morphism:

$$w_0, w_1 = h(w_0), w_2 = h(w_1) = h^2(w_0), \ldots .$$

The generated language $L(G)$ consists of all words in the sequence.

Two *DOL* systems may generate the same language, although their sequences are different. Thus, one may speak of *sequence equivalence* and *language equivalence* problems for *DOL* systems. The problems are involved and remained celebrated open problems until they were finally shown to be decidable. Before that it was known how an algorithm for solving the sequence equivalence problem could be transformed into an algorithm for solving the language equivalence problem, and vice versa. Neither one of the transformations is obvious: two *DOL* systems can generate the same language in quite a different order. The main idea is the decomposition of sequences into parts having properties (in particular, growth in the number of occurrences of all letters) that make them distinguishable in both systems.

Theorem 4.5. *Any algorithm for solving the language equivalence problem for DOL systems yields an algorithm for solving the sequence equivalence problem for DOL systems, and vice versa.*

Proof. Assume first that we know an algorithm for solving the language equivalence problem. We show how we can decide for two given *DOL* systems

$$G = (\Sigma, g, u_0) \text{ and } H = (\Sigma, h, v_0)$$

whether or not $E(G) = E(H)$. We assume without loss of generality that the two systems have the same alphabet. (We can also assume that $u_0 = v_0$ because, otherwise, the sequences cannot be the same.)

Construct two new *DOL* systems

$$G_a = (\Sigma \cup \{a\}, g_a, au_0) \text{ and } H_a = (\Sigma \cup \{a\}, h_a, av_0).$$

Here $g_a(a) = h_a(a) = a^2$; for letters of Σ, the values of g_a and h_a coincide with those of g and h respectively. It follows that the sequences of G_a and H_a consist of the words

$$a^{2^i} u_i \text{ and } a^{2^i} v_i, \quad i = 0, 1, \ldots$$

respectively. Clearly, $L(G_a) = L(H_a)$ exactly in case $E(G) = E(H)$. We can decide the validity of the former equation by our algorithm.

Conversely, assume that an algorithm for solving the sequence equivalence problem is known. Given two *DOL* systems G and H (in the same notation as above), we have to decide whether or not $L(G) = L(H)$. Since $L(G)$ is finite iff some word occurs twice in the sequence $E(G)$ and since the latter condition is easily decidable (we omit the details), we assume without loss of generality that both $L(G)$ and $L(H)$ are infinite.

We first explain the main idea. For any integers m and p ("initial mess" and "period"), we obtain from G the *DOL* system

$$G(p,m) = (\Sigma, g^p, g^m(u_0)) \, .$$

The sequence of $G(p,m)$ is obtained from $E(G)$ by omitting the initial mess and then taking every pth word only. In this way $L(G)$ is decomposed into a finite language and p *DOL* languages. The idea is to apply sequence equivalence testing for decompositions of $L(G)$ and $L(H)$ with different pairs (m,p). However, running systematically through all pairs will not constitute a good strategy. One has to be more selective, and use the results from earlier tests. Here Parikh vectors $\psi(w)$ will be a useful tool. A partial order $\leq$ for Parikh vectors of words over the same alphabet is defined by ordering the components. For words w and w', the notations $w \leq_P w'$ and $w <_P w'$ mean that $\psi(w) \leq \psi(w')$ and $\psi(w) < \psi(w')$ respectively. We need the following simple properties of Parikh vectors. (Recall that the *DOL* languages we are considering are infinite.)

(i) There are words with the property $w <_P w'$ in a *DOL* sequence. This can be easily seen by direct argument. It also follows by the so-called König Lemma.
(ii) There are no words w_i and w_j, $i < j$, in a *DOL* sequence such that $w_j \leq_P w_i$. Otherwise, the language would be finite. Consequently, the sequence does not contain two words with the same Parikh vector.
(iii) If $w_i <_P w_j$, then $w_{i+n} <_P w_{j+n}$ for all n.
(iv) Assume that w_i and w'_i, $i \geq 0$, are *DOL* sequences over an alphabet with n letters and that $\psi(w_i) = \psi(w'_i)$, $0 \leq i \leq n$. Then the sequences are Parikh-equivalent, that is, $\psi(w_i) = \psi(w'_i)$ for all i. This follows by linear algebra, [Sa So].

We now give an algorithm for deciding whether or not $L(G) = L(H)$. The algorithm uses the parameter m (length of the initial mess.) Initially, we set $m = 0$.

Step 1. Find the smallest integer $q_1 > m$ for which the exists an integer p such that

$$u_{q_1-p} <_P u_{q_1} \text{ and } 1 \leq p \leq q_1 - m \, .$$

Let p_1 be the smallest among such integers p. Determine in the same way integers q_2 and p_2 for the sequence v_i generated by H.

Comment. By property (i), Step 1 can always be accomplished, and the resulting numbers are unique. Clearly, a DOL sequence results if m first words are removed from a DOL sequence.

Step 2. If the two finite languages

$$\{u_i | m \leq i < q_1\} \text{ and } \{v_i | m \leq i < q_2\}$$

are different, stop with the conclusion $L(G) \neq L(H)$.

Comment. The following argument shows that the conclusion is correct. Assume first that $m = 0$. If there is a word w belonging to, say, the first but not to the second language and still $L(G) = L(H)$, then $w = v_i$ for some $i \geq q_2$. By property (iii), $L(H)$ contains a word w' with $w' <_P w$. By the choice of q_1, w' must come after w in the sequence $E(G)$, which contradicts (ii). Assume that $m > 0$. Then we must have increased the value of m in Step 3, before which we have concluded that the languages of Step 2 coincide. The argument remains the same — only the first m words in the sequences have been removed.

Step 3. Since the two languages in Step 2 are the same and no repetitions can occur, we must have $q_1 = q_2$. Now the known algorithm for sequence equivalence is applied. Check whether or not $p_1 = p_2$ and there is a permutation π defined on the set of indices $\{0, 1, \ldots, p_{-}1\}$ such that

$$E(G(p_1, q_1 + j)) = E(H(p_1, q_1 + \pi(j))), \ 0 \leq j \leq p_1 - 1.$$

If the answer is "yes", stop with the conclusion $L(G) = L(H)$. If the answer is "no", take q_1 as the new value of m and start again from Step 1.

Comment. The conclusion is correct. When entering Step 3, we know that

$$\{u_i | 0 \leq i < q_1\} = \{v_i | 0 \leq i < q_1\}.$$

By the test performed in Step 3 we also have

$$\{u_i | i \geq q_1\} = \{v_i | i \geq q_1\},$$

since both languages can be decomposed into pairwise equivalent sequences.

We have shown that the answer we obtain, following Steps 1–3, is always the correct one. We still have to show that we always actually obtain an answer, that is, the procedure terminates. This is obvious if $L(G) \neq L(H)$. Since the parameter m becomes arbitrarily large, a word belonging to only one of the languages is eventually detected in Step 2.

Assume that $L(G) = L(H)$. We have to show that the equality is detected during some visit to Step 3. Observe first that the values of p_1 and p_2 are never increased during successive visits to Step 1: if one visit produced the pair (p_1, p_2) and the next visit the pair (p'_1, p'_2), then $p'_1 \leq p_1$ and $p'_2 \leq p_2$. This follows by property (iii) because we always choose the smallest possible p-value. To show that the procedure cannot loop by producing the same pair (p_1, p_2), we prove that the same pair cannot be defined at more than $n+1$ consecutive visits to Step 1, where n is the cardinality of the alphabet Σ.

Assume the contrary: the same pair (p_1, p_2) is defined at $n+2$ consecutive visits. If $p_1 \neq p_2$ then the larger of p_1 and p_2 has to be decreased at the next visit to Step 1, because of property (iii) and the fact that Step 2 was passed after the preceding visit. Hence, $p_1 = p_2$. The same argument shows that p_1 must assume the maximal value $q_1 - m$.

Let m be the initial mess at the first of the $n+2$ visits under consideration. We know that

$$\{u_i | m + jp_1 \leq i < m + (j+1)p_1\} = \{v_i | m + jp_1 \leq i < m + (j+1)p_1\},$$

for $0 \leq j \leq n$. Thus, beginning with u_m and v_m, the sequences $E(G)$ and $E(H)$ contain $n+1$ segments of length p_1 such that each of these segments in $E(H)$ is a permutation of the corresponding segment in $E(G)$. It always has to be the same permutation. Otherwise, again by property (iii), the value of p_1 will be decreased at the next visit to Step 1. This is illustrated by the following example with $m=0$ and $p_1 = 5$:

$E(G):\quad u_0 u_1 u_2 u_3 u_4 | u_5 u_6 u_7 u_8 u_9 | u_{10} u_{11} u_{12} u_{13} u_{14} | u_{15} u_{16} u_{17} u_{18} u_{19} | \ldots$

$E(H):\quad u_1 u_0 u_3 u_4 u_2 | u_6 u_5 u_8 u_9 u_7 | u_{11} u_{10} u_{13} u_{12} u_{14} | u_{16} u_{15} u_{18} u_{19} u_{17} | \ldots$

The permutation is the same in the first two segments, but different in the third segment. From $E(H)$ we see that $u_9 <_P u_{12}$, which causes the period to be at most $3(< 5)$ after u_9 when $E(G)$ is considered.

Consequently, by property (iv), there is a permutation π on $\{0, 1, \ldots, p_1 - 1\}$ such that the sequences

$$E(G(p_1, m+j)) \text{ and } E(H(p_1, m+\pi(j)))$$

are Parikh-equivalent, for $0 \leq j \leq p_1 - 1$. However, because termination did not occur in Step 3, for some j the two sequences are not equivalent. Thus, for some $u \in L(G)$ and $v \in L(H)$,

$$\psi(u) = \psi(v), \ u \neq v.$$

Since $L(G) = L(H)$, v must be in $L(G)$ and, consequently, $E(G)$ contains two words with the same Parikh vector. This contradicts property (ii). □

We consider, finally, different variants of a problem whose decidability is still open. The language-theoretic variants concern the length of words. We first list five variants of this problem. The variants **P1** and **P2** are called the problem of the *existence of 0 in a Z-rational sequence*. (See [SaSo] for an explanation concerning the terminology.) The variants **P3**–**P5** are called the problem of a *constant level in DOL sequences*.

P1. Given a square matrix M with integer entries, does there exist a $j \geq 1$ such that the number 0 appears in the upper right-hand corner of M^j ?

P2. Given a square matrix M, a row vector π and a column vector η, all with integer entries and of the same dimension, does the number 0 appear in the sequence defined by
$$n_i = \pi M^i \eta \ ?$$
P3. Given two DOL sequences u_i and v_i, $i \geq 0$, does there exist a j such that
$$|u_j| = |v_j| \ ?$$
P4. Same as **P3** but u_i and v_i are $PDOL$ sequences.

P5. Given a DOL sequence u_i, $i \geq 0$. Does there exist a j such that
$$|u_{j+1}| = |u_j| \ .$$

For each of **P1–P5**, we find the required number, provided it exists, simply by testing all values from the beginning. The difficulty lies in the fact that we do not know where to stop. In other words, the set of "yes" instances is recursively enumerable.

Clearly, **P1** is a subproblem of **P2**. Hence, the decidability of **P2** implies the decidability of **P1**. Similarly, **P4** is a subproblem of **P3**. On the other hand **P5** becomes decidable if $PDOL$ sequences are considered. This follows because, for $PDOL$ sequences, $|u_{j+1}| = |u_j|$ holds exactly in case the minimal alphabet of u_j consists of letters that are mapped into a single letter by the morphism. Since the minimal alphabets occur in DOL sequences in an almost periodic fashion, this condition is decidable.

Consider **P5**. Given a DOL system generating the sequence u_i, we consider the growth matrix M and initial vector π of the system. Define, further, the column vector
$$\eta = (M - I)\eta_1 \ ,$$
where I is the identity matrix and η_1 is the column vector consisting of 1s. Then
$$|u_{j+1}| - |u_j| = \pi M^{j+1} \eta_1 - \pi M^j \eta_1 = \pi M^j \eta \ .$$

Hence, the decidability of **P2** implies the decidability of **P5**. A slightly more complicated argument shows that the decidability of **P2** implies the decidability of **P3** and, hence, that of **P4**.

The implications in the other direction, as well as the reduction of **P2** to **P1**, are based on matrix constructions such as increasing the dimension. An interested reader can find the details needed for the proof of the following theorem in [SaSo].

Theorem 4.6. *The problems* **P1–P5** *are equivalent with respect to decidability.*

Although their decidability status is open, it is generally conjectured that problems **P1–P5** are decidable. The problems are of "numerical" character. However, it is interesting to note that their decidability also implies the decidability of many

Problems with many faces 145

problems of a language-theoretic character. We mention without proof the following, rather old result [Ruohonen].

The *ultimate sequence equivalence problem* for DOL systems consists of deciding for two given DOL systems G and H, whether or not $E(G)$ and $E(H)$ differ from each other in only a finite number of terms.

Theorem 4.7. Any algorithm for one of problems **P1**–**P5** can be converted into an algorithm for solving the ultimate sequence equivalence problem for DOL systems.

In fact, it is known that the ultimate sequence equivalence problem is decidable. Thus, problems **P1**–**P5** appear to be more difficult than the ultimate sequence equivalence problem. We give, finally, an example of a problem whose decidability implies the decidability of problems **P1**–**P5**.

Thin languages were defined in Section 2.4. We say that a language is *properly thin* iff it contains no two words of the same length. Thus, contrary to the definition of thinness, no exceptional lengths are allowed.

Theorem 4.8. Any algorithm for deciding whether or not a given DOL language is properly thin yields an algorithm for all problems **P1**–**P5**.

Proof. By Theorem 4.6, it suffices to construct an algorithm for *P5*. Consider an arbitrary DOL system $G = (\Sigma, g, u_0)$. Let b be a letter not in Σ and define

$$\Sigma' = \{a' | a \in \Sigma\}.$$

"Primed versions" w' of words w are obtained by providing all letters with primes. (By definition, $\lambda' = \lambda$.) Define, further,

$$m = \max\{|u_0|, |g(a)| \mid a \in \Sigma\} + 1.$$

Consider the DOL system

$$G' = (\Sigma \cup \{b, b'\} \cup \Sigma', g', b^m u_0),$$

with the morphism g' defined by

$$g' : b \to b', b' \to b^m, a \to (h(a))', a' \to a \ (a \in \Sigma).$$

The first few words in the sequence $E(G')$ are

$$b^m u_0, b'^m u_1', b^{m^2} u_1, b'^{m^2} u_2', b^{m^3} u_2, b'^{m^3} u_3', \ldots.$$

Clearly, $L(G')$ is properly thin iff the answer is "no" to **P5** as regards G. The growth in the number of occurrences of b guarantees that $L(G')$ contains two words of the same length only if $E(G)$ contains two consecutive words of the same length. □

It is not known whether the converse of Theorem 4.8 holds. To use an algorithm for **P5** to decide proper thinness would require some upper bound concerning how far apart two words of the same length can lie in a DOL sequence. Rather surprising examples can be given in this respect.

4.2 Borderline between decidable and undecidable

Every subproblem of a decidable problem is itself decidable. This is clear because any algorithm for solving the whole problem can be applied to solve the subproblem. Conversely, the undecidability of the subproblem implies the undecidability of the whole problem: if the latter were decidable, so would be the subproblem, which contradicts the assumption.

This suggests the following line of thought. Assume that we start with, as is often the case, an undecidable problem and a decidable subproblem. We then restrict the former, still retaining undecidability, and/or extend the latter, still retaining decidability. In this fashion we try to approximate the borderline between decidability and undecidability. In doing so, we have in mind some parameter indicating the size of the problem, and we are sometimes able to determine a cut-point value of the parameter, where the problem changes from decidable to undecidable. How informative the knowledge of such a cut-point value will be depends essentially on the nature of the parameter. Certain parameters tell rather little about the borderline, and other measures are called for to obtain a sharper view.

A very typical example is the Post Correspondence Problem. As in Chapter 2, consider instances $PCP = (g, h)$, where Σ and Δ are the range and the domain alphabets of the morphisms g and h. An increasing sequence of subproblems of the Post Correspondence Problem results if we consider only instances PCP, where the cardinality of Δ does not exceed k, for $k = 1, 2, \ldots$. The whole Post Correspondence Problem is of course the union of all of these subproblems. It was seen in Examples 2.6 and 2.7 that the cut-point value of the parameter k is between 1 and 2: the subproblem with $k = 1$ is decidable, whereas the subproblem with $k = 2$ is undecidable. This result is easily obtained and says very little. It only indicates that two letters suffice to encode an arbitrary number of letters in a uniquely decodable way.

In the same way we can define an increasing sequence of subproblems using the cardinality of the domain alphabet Σ as the parameter. The exact cut-point value is not known but Theorem 2.3 tells us that the value lies between 2 and 9. Even if the exact value were known, it is questionable whether knowledge of it would capture something essential about the undecidability of the Post Correspondence Problem.

Let us take a somewhat different illustration. Consider a rewriting system (Σ, P) and two words w and w' over Σ. We ask whether or not w yields w', that is,

$$w \Rightarrow^* w',$$

according to the rewriting rules in P. This problem is referred to as the *semi-Thue problem for semigroups*. If P is symmetric, that is, whenever $u \to v$ is in P then so is $v \to u$, we speak of the *Thue problem for semigroups*. In this case we may view the rewriting rules as equations $u = v$, and the yield relation indicates that the subwords u and v are interchangeable. The Thue problem is also often referred to as the *word problem*. Indeed, in algebraic terminology, the equations in P constitute the defining relations for a semigroup, and the word problem amounts to asking whether or not two words define the same element of the semigroup.

Semi-Thue and Thue problems can also be defined for *groups*. From the point of view of rewriting, this means that the letters of Σ appear in pairs (a, a^{-1}) such that both aa^{-1} and $a^{-1}a$ can be rewritten as λ, and vice versa. Finally, all problems mentioned can be formulated both for the *commutative* (here we prefer this term to "Abelian") and general (noncommutative) case.

In this fashion we have defined three binary parameters: semi-Thue or Thue, semigroup or group, general or commutative. The values of the parameters can be fixed independently of each other, and each choice of parameter values specifies a problem. We thus have eight different problems. Which ones are decidable, and which are undecidable? Where does the borderline go?

To avoid misunderstandings, let us define one of the problems, say the *Thue problem for commutative semigroups*, very explicitly. Let (Σ, P) be a rewriting system, where P contains all productions of the form $ab \to ba$, where a and b are letters of Σ, and whenever $u \to v$ is in P then so is $v \to u$. The Thue problem for commutative semigroups consists of finding an algorithm which, after receiving such a rewriting system and two words w and w' as its input, determines whether or not $w \Rightarrow^* w'$. In this formulation we present matters in terms of a rewriting system. In algebraic terms the problem amounts to the word problem for commutative semigroups (specified by the finite set P of defining relations). Here we formulated the *uniform* version of the problem: we require an algorithm that settles the question for all commutative semigroups and all pairs of words. Of course, we can also consider the word problem for one particular semigroup.

It turns out that in this discrete setup with eight problems the borderline goes along commutativity. All four "commutative problems" are decidable, whereas the remaining four problems are undecidable. To reach this conclusion, it of course suffices to establish the claims for the largest commutative and the smallest noncommutative problem. In other words, one has to prove that the Thue problem (also referred to as the word problem) for groups is undecidable but the semi-Thue problem for commutative semigroups is decidable. Various versions of the latter problem were considered in Section 4.1.

Thus, in this setup the borderline is known. Here we do not want to enter into the details of the history of the exploration of this territory. As expected, one did not come to the borderline at once. The decidability of smaller problems and the undecidability of larger problems was established before the proofs for the problems considered in the preceding paragraph.

Similarly as the Post Correspondence Problem, Hilbert's Tenth Problem can also be divided in many ways into an increasing sequence of subproblems. We leave it to the reader to consider various ways for doing this, keeping in mind the issue of the borderline.

Problems for recursively enumerable languages can be divided in a very natural way into an increasing sequence of subproblems. One just considers increasing hierarchies of language families, and restricts the problem for each family.

By considering the Chomsky hierarchy, one obtains in this fashion a rough idea about the borderline between decidability and undecidability for a specific problem. We will discuss some very typical problems briefly. We refer to the families in the Chomsky hierarchy using numbers 0–3 in the customary way. In other words, recursively enumerable languages are referred to as type 0 languages, context-sensitive as type 1, context-free as type 2, and regular as type 3 languages.

It should now be clear what we mean by saying that the borderline lies between 0 and 1. This is the case for the *membership problem*. It is decidable for context-sensitive languages but undecidable for recursively enumerable languages. Both results are, in fact, a bit stronger. The decidability is *uniform*: there is an algorithm that decides for all pairs (G, w), where G is a context-sensitive grammar and w a word, the question of w belonging to $L(G)$. The undecidability concerns single grammars: there are specific grammars G such that no algorithm settles the question of a given word w belonging to $L(G)$.

As regards the *emptiness* and *finiteness problems*, the borderline lies between 1 and 2. There is an algorithm for deciding whether or not a given context-free language is empty or finite, whereas both problems are undecidable for context-sensitive languages. The borderline lies between 2 and 3 for the following problem of *universality*. Given a grammar G with the terminal alphabet Σ, one has to decide whether or not $L(G) = \Sigma^*$.

Thus, it is easy to obtain examples of borderlines *between* any two families in the Chomsky hierarchy. What about problems that are decidable or undecidable for all families? By Rice's Theorem, every nontrivial property is undecidable for recursively enumerable languages — thus the borderline between 0 and 1 is an extreme one. One can also say that the borderline between 3 and 2 is an extreme one. All natural properties are decidable for regular languages. The problems known to be undecidable for regular languages have the flavour that something nonregular is hidden in the property defining the problem.

Of special importance and interest in language theory is the *problem of equivalence*. Given two devices generating, accepting or otherwise defining a language, do the two languages thus obtained coincide? A specific decision problem results when the devices are taken from a specific class, such as the class of context-free grammars. Earlier (for instance, see Theorem 2.12) we illustrated the use of the Post Correspondence Problem in such cases.

Considering only the families in the Chomsky hierarchy, the borderline for the equivalence problem lies between 2 and 3. However, this gives very little information, since the gap is huge. A celebrated open problem concerns *deterministic context-free languages*: is their equivalence problem decidable? Thus, we do not know on which side of the borderline this specific problem lies.

Better approximations of the borderline for the equivalence problem are obtained by considering *comparative equivalence problems*. This means that the two languages come from different families. Thus, the comparative equivalence problem between regular and linear languages consists of finding an algorithm that settles the validity of the equation $L(G_R) = L(G_L)$ for given regular and linear grammars G_R and G_L. We know from Theorem 2.12 that this problem is undecidable. We express this briefly by saying that the equivalence problem for (**REG,LIN**) is undecidable. This result is clearly much stronger than the undecidability of the equivalence problem for (**CF,CF**).

We use the further abbreviations **DET** and **UNAMB** for the families of deterministic and unambiguous (context-free) languages. Comparative equivalence problems shed some light on the open equivalence problem for deterministic languages, that is, on the equivalence problem for (**DET,DET**). In fact, a decidable problem results even if one of the families is made larger, provided the other family is simultaneously made smaller.

Theorem 4.9. The equivalence problem for (**REG,UNAMB**) is decidable.

The rather complicated proof of Theorem 4.9 is based on formal power series, see [Ku Sa]. The proof for the subproblem defined by (**REG,DET**) is easy because both **REG** and **DET** are (effectively) closed under complementation and intersection with regular languages. Hence, the symmetric difference between a regular language and a deterministic language can be constructed and its emptiness decided.

Thus, the best results from the above discussion are: the equivalence problem for (**REG,LIN**) is undecidable, whereas the equivalence problem for (**REG,UNAMB**) is decidable. Other, more specialized, results about comparative equivalence problems relevant to the one concerning (**DET,DET**) are known.

As regards comparative equivalence problems involving language families generated by L systems, the basic undecidability results for (**OL,OL**) and (**CF,OL**) should be contrasted with the decidability of the equivalence problems for (**OL,DOL**), (**CF,DOL**) and (**REG,OL**). For details, see [Ro Sa].

To summarize the results concerning comparative equivalence problems with these families, the borderline is approximated from the side of undecidability by the pairs (**REG,LIN**), (**OL,OL**) and (**CF,OL**). Observe that these three problems are pairwise incomparable as regards inclusion. From the side of decidability the borderline is approximated by the pairs (**REG,UNAMB**), (**OL,DOL**), (**CF,DOL**) and (**REG,OL**), also giving rise to pairwise incomparable problems.

We conclude this section with illustrations from very recent language-theoretic works, where the reader might wish to try to investigate the borderline that is rather inadequately understood at the time of writing.

We define the notion of a *pattern language*. Let Σ and V be disjoint alphabets, *terminals* and *variables*. For simplicity, we assume that Σ is the binary alphabet: $\Sigma = \{0,1\}$. Any nonempty word over the alphabet $\Sigma \cup V$ is called a *pattern*. A pattern α generates a language $L(\alpha)$ over Σ, consisting of all words obtained from α by uniformly substituting the variables in α with nonempty words over Σ. Formally,

$$L(\alpha) = \{h(\alpha) \mid h : (\Sigma \cup V)^+ \to \Sigma^+ \text{ is a morphism with } h(0) = 0, h(1) = 1\}.$$

For instance,

$$L(x0x) = \{w0w \mid w \in \Sigma^+\},$$
$$L(x1y0x) = \{w1u0w \mid w, u \in \Sigma^+\}.$$

both being noncontext-free languages.

We say that a pattern β results from a pattern α by *renaming the variables* iff there is a one-to-one mapping φ from the set of variables in α on to that in β such that β equals the pattern obtained from α by replacing every variable x by $\varphi(x)$. For instance, $u0uu1zyz0y$ results from $x0xx1yzy0z$ by renaming the variables.

Theorem 4.10. The equation $L(\alpha) = L(\beta)$ holds between patterns α and β iff β results from α by renaming the variables. Consequently, the equivalence problem is decidable for pattern languages.

Proof. Clearly, the first sentence yields an algorithm and, hence, the second sentence follows from the first. The "if"-part of the first sentence is obvious. Conversely, assume that $L(\alpha) = L(\beta)$. The claim now follows from the formal definition of $L(\alpha)$, for instance by scanning α from left to right. If the first letter is a variable, the first letter of β is also a variable. In this case we rename the first letter of α accordingly. If the first letter of α is a terminal, the first letter of β is the same terminal. Proceeding inductively, we assume that we have already been able to rename the first k letters of α in such a way that they coincide with those of β. Then the equation $L(\alpha) = L(\beta)$ implies that the $(k+1)$th letter is, in both α and β, either the same terminal or a variable that does not contradict the renaming performed so far. (In other words, either the variables did not occur before or occurred in the same positions in both α and β.) The equation also implies that α and β are of the same length.

□

A problem that is closely related to the equivalence problem is the *inclusion problem*: we look for an algorithm to decide whether or not $L_1 \subseteq L_2$ holds for given languages L_1 and L_2 from a specified family. Clearly, the decidability of the inclusion problem always implies the decidability of the equivalence problem; equality can be found by testing inclusion in both directions. The reverse implication is not valid.

Indeed, in spite of the straightforward decidability of the equivalence problem, the inclusion problem has recently been shown to be undecidable. The following example indicates some of the difficulties involved.

Example 4.5. Consider the pattern $yxxz$. We claim that

(∗) $$\Sigma^6 \Sigma^* \subseteq L(yxxz).$$

Indeed, let w be an arbitrary word of length ≥ 6, and let a_i be the ith letter of w, for $i = 1, 2, \ldots, k, k \geq 6$. We assume without loss of generality that $a_2 = 0$, the case $a_2 = 1$ being symmetric. Depending on the values of $a_3 - a_5$, we choose the words x, y and z as follows:

For $\quad a_3 = 0$, choose $y = a_1$, $x = 0$, $z = a_4 \ldots a_k$.
For $\quad a_3 = a_4 = 1$, choose $y = a_1 a_2$, $x = 1$, $z = a_5 \ldots a_k$.
For $\quad a_3 = 1$, $a_4 = a_5 = 0$, choose $y = a_1 a_2 a_3$, $x = 0$, $z = a_6 \ldots a_k$.
For $\quad a_3 = 1$, $a_4 = 0$, $a_5 = 1$, choose $y = a_1$, $x = 01$, $z = a_6 \ldots a_k$.

Hence, in all cases, w is in $L(yxxz)$, and (*) follows.

If α is any pattern with $|\alpha| \geq 6$, then $L(\alpha) \subseteq \Sigma^6 \Sigma^*$ and hence, by (*),

$$L(\alpha) \subseteq L(yxxz).$$

However, for some αs, this inclusion is difficult to establish directly. In particular, it is not possible to generalize the idea of Theorem 4.10 and factorize α into four parts corresponding to y, x, x, z. For instance, what would the parts be for $\alpha = uu0u1u$? But nevertheless

$$L(uu0u1u) \subseteq L(yxxz).$$

$\square$

We make another excursion to recent language theory. We call a language F the *deletion language* defined by (w, L), where w is a word and L is a context-free language, iff F results from w by deleting subwords belonging to L, that is,

$$F = \{w_1 w_2 \mid w = w_1 x w_2, \ x \in L\}.$$

(The words involved may also be empty.)

Clearly, every deletion language is finite. Not every finite language is a deletion language. If $\{a, b, c\}$ were a deletion language, then each of the three letters would appear either as a prefix or as a suffix in w, which is possible only for two of them. The language

$$\{w \in \{a, b\}^* \mid |w| \leq 2\}$$

is not a deletion language, but with the omission of the word ba it becomes one. We leave the construction of a simple algorithm for deciding whether or not a given finite language is a deletion language to the reader.

Thus, decision is easy if we let both w and L vary. Things become trickier if we keep L and F fixed and ask whether or not there exists a w such that F is a deletion language defined by (w, L).

Theorem 4.11. There is no algorithm which decides, for a given finite language F and context-free language L, whether or not, for some word w, F is the deletion language defined by (w, L).

Proof. We apply reduction to the Post Correspondence Problem in very much the same way as in Section 2.4. Consider an instance PCP defined by the lists

$$(\alpha_1, \ldots, \alpha_n) \text{ and } (\beta_1, \ldots, \beta_n),$$

where the alphabet of the αs and βs does not contain the letters a, b, c, d (needed for other purposes). The context-free grammar G determined by the productions

$$S \to cS_\alpha \, , \, S \to S_\beta d \, ,$$

$$S_\alpha \to ba^i S_\alpha \alpha_i \, , \, S_\alpha \to ba^i \alpha_i \, , \, i = 1, \ldots, n \, ,$$

$$S_\beta \to ba^i S_\beta \beta_i \, , \, S_\beta \to ba^i \beta_i \, , \, i = 1, \ldots, n \, ,$$

generates the language

$$\begin{aligned} L(\alpha, \beta) & = \{ cba^{i_1} \ldots ba^{i_k} \alpha_{i_k} \ldots \alpha_{i_1} \mid k \geq 1, \, 1 \leq i_j \leq n \} \\ & \cup \; \{ ba^{i_1} \ldots ba^{i_k} \beta_{i_k} \ldots \beta_{i_1} d \mid k \geq 1, \, 1 \leq i_j \leq n \} \, . \end{aligned}$$

Consider the language $F = \{c, d\}$. There is a word w such that F is the deletion language defined by $(w, L(\alpha, \beta))$ exactly in case the instance PCP possesses a solution. This follows because such a w is necessarily of the form $w = cw_1 d$, where w_1 is associated with a solution. □

Actually, the proof above gives a stronger result than is claimed in the statement of the theorem. The claim concerns the nonexistence of an algorithm that is uniform for pairs (F, L). The proof shows that there is not even an algorithm for pairs $(\{c, d\}, L)$. Thus, there is a fixed finite language, namely $F = \{c, d\}$, such that there is no algorithm which decides, for a given L, whether, for some w, F is the deletion language defined by (w, L). On the other hand, there are also finite languages F for which the problem mentioned is decidable. An obvious example is $F = \{\lambda\}$. Then, for a given L, a word w exists as required iff L is nonempty. However, for all other finite languages F, the problem is undecidable. Thus, here we have a borderline between undecidable and decidable, crossing through the family of finite languages.

4.3 Mathematical proofs: step by step and zero-knowledge

A formal mathematical proof proceeds step by step. There are starting points, axioms and rules of inference. A step in the proof consists of applying a rule of inference to something already available, that is, an axiom or a previous stage in the proof. Of course, this idealized situation is never reached in practice. However, it is the situation that one has in mind when one speaks about formally undecidable propositions.

One purpose in giving a proof is to convince another party about a certain state of affairs, such as having information of something. In other words, one party convinces another party that it is in the possession of a certain knowledge. By modern cryptographic techniques, it is possible to do this without revealing any

Mathematical proofs

of the knowledge itself. Such *zero-knowledge proofs* have opened new vistas for communication. For instance, a mathematician can convince another mathematician that he has settled a conjecture without giving any details, not even whether or not the conjecture is true. Or one can prove knowledge of a password without revealing a bit of the password itself. In some sense such results are contraintuitive. However, they undoubtedly constitute an important new aspect concerning mathematical proofs in general. Therefore, we want to make a short excursion into zero-knowledge proofs, although this topic is somewhat removed from the mainstream of this book.

Much of the subsequent discussion deals with notions from the theory of computational complexity. In particular, we speak of (computationally) *intractable problems* and *one-way functions*, already discussed in Section 1.2.

We will consider the following situation. The *prover* (denoted by P and also called Peter) knows some information, such as the prime factorization of a large integer (this is regarded in general as intractable), the proof of a long-standing conjecture such as Fermat's last problem, a password or an identification number. The prover would like to convince the *verifier* (V, Vera) beyond any reasonable doubt that he possesses this information.

The simple way for P to convince V is just to disclose the information, that is, to present the prime factors, the proof, the password or the identification number. Then V can check the correctness herself, for instance by multiplying the prime factors and verifying that the result is the original large integer. This is a *maximum disclosure proof*, where V actually learns the information and can later show it to someone else, and even claim that she determined the factorization herself.

In a *minimum disclosure proof* P convinces V that he has the information but this happens in a way that does not reveal any of the information and, thus, does not in any way help V to find out the information. V can make the probability of P cheating arbitrarily small. For instance, P could only pretend to know the prime factors. Thus, V will be convinced that P actually knows the factors but she does not learn the factors herself and cannot state anything about them to a third party.

Example 4.6. We present a simple minimum disclosure proof about knowledge of the factors of $n = pq$, where p and q are large prime numbers, say, one hundred digits each. Thus, n is known to both P and V, and P wants to convince V that he also knows p and q. (Of course, this is computationally equivalent to knowing p because division is computationally easy.)

Step 1. V chooses a random integer x and tells x^4, reduced modulo n, to P.

Step 2. P tells $x^2 \pmod{n}$ to V.

Observe, first, that V obtains no information that is new to her because she can square x herself. Why does P's answer (provided it is correct) convince V about P's knowledge of p and q? This follows because extracting square roots is computationally equivalent to factoring n. Let us consider this in more detail.

It is a number-theoretic fact that each positive integer a has at most four square roots incongruent modulo n. Let us denote them by $\pm b$ and $\pm c$. In the simple case where $n = 15$, the square roots of 4 are ± 2 and ± 7, also representable in the form $2, 13, 7, 8$. (Since we are working modulo 15, the numbers -2 and 13 are considered to be equal.) The number 12, for instance, has no square roots. Assume now that, for some a, P is able to compute the different square roots b and c. Since $b^2 - c^2 = (b+c)(b-c) \equiv 0 \pmod{n}$ and $b \not\equiv \pm c \pmod{n}$, one of the factors p and q divides $b + c$ and the other $b - c$. Thus, P finds p or q by computing the greatest common divisor of n and $b + c$, which is computationally easy by Euclid's algorithm.

In order to pass Step 2, P not only has to extract a square root of x^4 but the particular one among the square roots that is a square itself. If he succeeds in this, he can also factor n, as seen above. It is worth mentioning that without knowing p and q, it is intractable to recognize numbers of the form x^2.

The small possibility of P succeeding by mere luck can be made still smaller by repeating Steps 1–2 with different values of x. However, in this protocol one round essentially suffices.

□

Let us repeat our basic setup and requirements for minimum disclosure proofs. Assume that the information we are talking about is a proof of a theorem.

(i) If the prover does not know a proof, his chances of convincing the verifier that he knows a proof are negligible.
(ii) The verifier does not get the slightest hint of the proof, apart from the fact that the prover knows a proof. In particular, the verifier cannot prove the theorem to anyone else without proving it herself from scratch.

Minimum disclosure proofs are conceivable even if the prover has no definite proof to start with but only an argument that is very likely to be true. For instance, P might be quite convinced that $n = pq$ is the prime factorization, although p can be further factorized. P can transfer his conviction to V in a minimum disclosure manner.

The protocol in Example 4.6 was constructed in a very *ad hoc* manner, based on the special interconnection between factoring and extracting square roots. We now describe in informal terms a useful general idea, that of a *lockable box*. The verifier cannot open it because the prover has the key. On the other hand, the prover has to *commit himself* to the contents of the box, that is, he cannot change the contents when he opens the box. In fact, the verifier may watch when the prover opens the box.

In more formal terms, the hardware of the lockable boxes is replaced by one-way functions. Locking the information x into a box means applying a one-way function f to x. V can now handle $f(x)$ without knowing what x is, since V is not in the possession of the inverse f^{-1}. On the other hand, P cannot change $f(x)$, since V knows it. Under the further assumption that f is injective, this means that P

cannot change x, that is, the contents of the box. When P opens the box and gives x to V, she can immediately verify that $f(x)$ is the number she had before.

Example 4.7. Here we discuss a one-way function possible for the construction of lockable boxes. We assume that each box will contain only one bit. Boxes containing more information can be replaced by several one-bit boxes that are opened simultaneously.

The method described below is based on the (currently generally accepted) assumption that the computation of *discrete logarithms* is intractable. Let p be a large prime number. There are *generators* g such that all numbers $1, \ldots, p-1$ are congruent to a power $g^i \pmod{p}$. If $j \equiv g^i \pmod{p}$, i is called the *discrete logarithm* of j to base g. In the simple case $p = 7$, $g = 3$, we have the table

Number	1	2	3	4	5	6
Logarithm	6	2	1	4	5	3

In this case the number 2 is not a generator: only the numbers 1, 2, 4 have a discrete logarithm to base 2.

Thus, it is computationally intractable to find i from $g^i = j$, whereas the exponentiation $g^i = j$ is easy when i is given. Both g and p are considered to be fixed during the following discussion. This means that P and V have agreed about g and p. Below it is understood that numbers are reduced modulo p, that is, the smallest nonnegative remainder of the number is taken. Consequently, equations are, in fact, congruences. The following protocol is used when P wants to lock a bit b ($b = 0$ or $b = 1$) into a box. It is not important how the random numbers (that are actually only pseudo-random in view of Chapter 5) are generated.

Step 1. V tells P a random number r, $1 < r < p - 1$.

Step 2. P chooses a random number x and tells V the "box" $y = r^b g^x$.

Step 3. To open the box, P gives V the "key", that is, P gives x to V.

After Step 3, V knows b because $b = 0$ iff $y = g^x$. On the other hand, x does not help V to open any other boxes. By knowing just y (and r), V has no information about b because all numbers in question are of both of the forms g^x and rg^x.

Let us see why the protocol forces P to commit himself to the bit b, that is, he cannot change b after Step 2. Suppose the contrary: P is able to choose two numbers x and x' in Step 2 such that

$$g^x = rg^{x'} = y.$$

When opening the box, P could then give x or x' as the key, depending on whether he wants the bit $b = 0$ or $b = 1$ to appear in the box. But because $g^{x-x'} = r$, this would mean that P has been able to compute the discrete logarithm of r, which is a number chosen by V.

□

We use the term *"zero-knowledge proof"* synonymously with the term "minimum disclosure proof". Sometimes, depending on the formulation, there are slight differences between these two notions. Such differences are of no concern to us in our informal presentation. In a zero-knowledge proof V becomes convinced about P's knowledge but obtains no knowledge that is new to herself. A good way to visualize this is to think how V could obtain all the information she obtains in the course of the protocol *entirely without* P, that is, how V could play the protocol as a solitaire game, without P participating at all. This condition has to be satisfied for the proof to be zero-knowledge.

We now present a zero-knowledge protocol for the *satisfiability problem for propositional formulas*. The great importance and versatility of this problem will also be explained below. We begin with a description of the problem.

We consider (well-formed) formulas α built from propositional *variables* $x_1, x_2, \ldots$, by the use of *connectives* $\sim, \vee, \wedge$ (negation, disjunction, conjunction). A *truth-value assignment* for such a formula α is a mapping f of the set of variables occurring in α into the set $\{T, F\}$. For any given truth-value assignment, the truth-value assumed by α can be computed using the truth-tables of the connectives. The computation can also be done quickly. A formula α is *satisfiable* if it assumes the value T for at least one truth-value assignment.

Example 4.8. Consider the formula

$$\alpha = (x_1 \vee \sim x_2 \vee x_3) \wedge (x_2 \vee x_3) \wedge (\sim x_1 \vee x_3) \wedge \sim x_3 \,.$$

Assume that α is satisfiable. Then, for some assignment, each of the four conjunctive clauses assumes the value T. The last clause forces the assignment $x_3 = F$. Hence, by the third clause, $x_1 = F$, and by the second clause $x_2 = T$. But for this assignment the first clause assumes the value F. This contradiction shows that α is not satisfiable.

□

In special cases such as the one above, satisfiability can be checked by an *ad hoc* method. However, there is no general method better than exhaustive search, that is, search through all possible 2^k truth-value assignments, where k is the number of variables. This makes the task computationally intractable. It is already computationally infeasible in the case of a few hundred variables. The satisfiability problem is NP-complete. Indeed, it is very basic among NP-complete problems in the sense that it constitutes perhaps the most suitable reference point for NP-complete problems: many problems can be shown to be NP-complete by reduction to the satisfiability problem.

We now show how a zero-knowledge proof can be constructed for the satisfiability problem. The notions should be sufficiently clear so that we can present the result as a formal theorem. Thus, the basic setup is as follows. Both the prover P and the verifier V know a propositional formula α. Besides, P knows a truth-value assignment f satisfying α. P convinces V about his knowledge of f without revealing anything concerning f. More explicitly, V becomes convinced that P is not

Mathematical proofs

cheating with a probability that can be made arbitrarily high, and all information V might eventually obtain concerning f she could also obtain without P.

For simplicity, we restrict our attention to propositional formulas in a specific form, 3-*conjunctive normal form*: a conjunction of disjunctions, where each disjunction consists of three literals. By a *literal* we mean either a variable or its negation. As far as complexity is concerned, this restriction is not essential. The satisfiability problem for propositional formulas in 3-conjunctive normal form is NP-complete.

Theorem 4.12. The satisfiability problem for propositional formulas in 3-conjunctive normal form possesses a zero-knowledge proof.

Proof. We will first present the protocol. It will then be illustrated by an example. Finally, the conditions for zero-knowledge will be discussed.

Thus, P and V know a propositional formula α. Assume that α has r propositional variables and s disjunctive clauses in the conjunction. We are now ready to describe the protocol.

Step 1. P prepares and gives V three sets of locked boxes, referred to as *variable*, *truth-value* and *assignment boxes*. Variable and truth-value boxes, VAR_i and TV_i, $i = 1, \ldots, 2r$, correspond to pairs (x, y), where x is a variable and y is a truth-value (T or F). The number of such pairs is $2r$. For each pair (x, y), there is an i such that x is locked in VAR_i and y in TV_i, but the ordering of the pairs is random. The indices of the assignment boxes $ASSIGN_{i,j,k}$ range from 1 to $2r$ and from ~ 1 to $\sim 2r$. Thus, the number of assignment boxes is $(4r)^3$. (The total number is smaller if it is assumed that no variable appears twice in a clause, disjunctions are ordered, etc.) Each of them contains the number 0 or 1 according to the following rule. Consider a disjunctive clause $\beta = x \vee y \vee z$, where x, y, z are literals. Assume that $x = x_{i'}$ and that i'' is the index such that $x_{i'}$ is in the box $VAR_{i''}$ and $f(x_{i'})$ is in the box $TV_{i''}$. Then we define $i = i''$. If $x =\sim x_{i'}$ and i'' is the index such that $x_{i'}$ is in the box $VAR_{i''}$ and $f(x_{i'})$ is in the box $TV_{i''}$, we define $i =\sim i''$. The values j and k are defined similarly, starting from y and z. Under these circumstances, the number 1 is locked in the box $ASSIGN_{i,j,k}$. All other assignment boxes contain the number 0. Thus, altogether, s assignment boxes contain the number 1.

Step 2. V gives one of the two commands "truth" or "consequences" to P.

Step 3. If V's command was "consequences", P opens all variable and assignment boxes. If her command was "truth", P opens all truth-value boxes and, moreover, all those assignment boxes $ASSIGN_{i,j,k}$ where each of the three indices is either of the form x with F in the truth-value box TV_x, or of the form $\sim x$ with T in TV_x.

Step 4. V fails P either if in the case of "truth" the number 1 appears in some of the opened assignment boxes, or else in the case of "consequences" the assignment boxes containing the number 1 do not yield the original formula α.

Step 5. If P passed in Step 4, V either accepts him (that is, V is convinced) or requests another round of the protocol, depending on whether or not she has already reached her preset confidence level.

Before commenting on the protocol and concluding the proof, we look at an example.

Example 4.9. Consider the following propositional formula α in 3-conjunctive normal form having five variables and eleven disjunctive clauses:

$$(\sim x_1 \vee \sim x_2 \vee \sim x_3) \wedge (x_1 \vee x_2 \vee \sim x_4) \wedge (\sim x_1 \vee x_2 \vee x_4)$$

$$\wedge (x_1 \vee x_2 \vee \sim x_5) \wedge (x_1 \vee \sim x_2 \vee \sim x_5) \wedge (x_1 \vee x_3 \vee x_4)$$

$$\wedge (\sim x_1 \vee x_3 \vee \sim x_5) \wedge (x_1 \vee \sim x_4 \vee x_5) \wedge (x_2 \vee \sim x_3 \vee x_4)$$

$$\wedge (x_3 \vee x_4 \vee x_5) \wedge (x_3 \vee \sim x_4 \vee x_5) \, .$$

The following general observation should first be made. A formula with only few disjunctive clauses is satisfiable, and so there is nothing to prove. For five variables, there are altogether $2^5 = 32$ possible truth-value combinations. Each clause may exclude four of them. For instance, the clause $x_1 \vee x_2 \vee x_3$ excludes the combinations

$$FFFTT, \; FFFTF, \; FFFFT, \; FFFFF$$

for the five variables x_1, x_2, x_3, x_4, x_5: for each of the resulting truth-value assignments the formula necessarily assumes the value F. Thus, a formula having at most seven clauses is "automatically" satisfiable because at most 28 truth-value combinations are excluded. This is only an upper bound because there may be same truth-value combinations among the combinations excluded by two different clauses. For instance, the conjunction $(x_1 \vee x_2 \vee x_3) \wedge (x_1 \vee x_4 \vee x_5)$ excludes only seven combinations, since the combination $FFFFF$ is excluded by both of the clauses.

Since our α possesses eleven clauses, satisfiability cannot be established by the above simple counting argument. However, α is satisfiable and P knows a truth-value assignment f with

$$f(x_2) = f(x_3) = T, \; f(x_1) = f(x_4) = f(x_5) = F,$$

for which α assumes the value T. (In fact, this is the only possible assignment but this state of affairs is not important in our considerations.) P wants to convince V about his knowledge of f without revealing anything about f.

To follow the protocol, P still has to select a random order for the pairs (x, y), where x is a variable and y is a truth-value. The order he selects is given in the subsequent table of the contents of the variable and truth-value boxes:

Mathematical proofs

i	VAR_i	TV_i
1	x_4	F
2	x_1	T
3	x_5	F
4	x_2	F
5	x_2	T
6	x_3	T
7	x_1	F
8	x_5	T
9	x_3	F
10	x_4	T

The assignment boxes are no less than 8000 in number. (If we assume that the indices appear in increasing order of magnitude and no index, negated or nonnegated, appears twice, then the smaller total number 960 suffices.) They all contain the number 0, except for the following eleven boxes which contain the number 1:

$$ASSIGN_{\sim 5,\sim 6,\sim 7}\,,\ ASSIGN_{\sim 1,5,7}\,,\ ASSIGN_{1,5,\sim 7}\,,$$
$$ASSIGN_{\sim 3,5,7}\,,\ ASSIGN_{\sim 3,\sim 5,7}\,,\ ASSIGN_{1,6,7}\,,$$
$$ASSIGN_{\sim 3,6,\sim 7}\,,\ ASSIGN_{\sim 1,3,7}\,,\ ASSIGN_{1,5,\sim 6}\,,$$
$$ASSIGN_{1,3,6}\,,\ ASSIGN_{\sim 1,3,6}\,.$$

The eleven boxes are listed above in the order obtained from the clauses of α. However, P gives all assignment boxes to V, say, in some alphabetic order.

The protocol can now be carried out. If V commands "consequences", she learns α (in a somewhat permuted form). But she already knew α. Nothing of f is revealed, since TV-boxes are not opened at all.

If V commands "truth", she learns in which boxes the Ts and Fs are but nothing about their connection with the variables. Moreover, she learns that the assignment f does not assign the value F to any of the clauses of α. This follows because 0 appears in all those 120 assignment boxes where all three indices come from the set $\{1, 3, 4, 7, 9, \sim 2, \sim 5, \sim 6, \sim 8, \sim 10\}$. Also, this does not reveal anything concerning the association of T and F with the individual variables.

Clearly, P fails in Step 4, either in the "truth" or "consequences" scenario if he doesn't know a correct truth-value assignment.

□

We now conclude the proof of Theorem 4.12. We have seen already in the example why V does not learn anything about f — the same argument holds in the general case as well. It is essential that, if another round of the protocol is called for, then everything is started from scratch. If something remains until the next round, say, from the random order, then consecutive "truth" and "consequences" commands may reveal facts about the truth-value assignment f.

If P guesses correctly what V's command in Step 2 will be, then he passes the whole round of the protocol. If the command will be "truth", then P locks the number 0 in all assignment boxes in Step 1. If it will be "consequences", then P just takes care of that the correct α will be found from the assignment boxes. If P does not know f, each round will decrease the probability of P not failing by the factor $\frac{1}{2}$. Thus, if V is satisfied with an error rate of less than one in a million, then twenty rounds suffice.

Observe, finally, that V can obtain the same information as in the protocol with P just by herself, that is, without P. She just plays the "guessing game" in Step 1. The opened boxes look exactly the same as in the protocol with P. The only difference is that V's conviction about P's knowledge is missing. □

The above argument can be used to convert any mathematical proof into a zero-knowledge proof. Suppose you know a proof of Fermat's last problem. There should be no "hand-waving" in the proof. The proof must have been formalized within some proof-theoretic system in sufficient detail for a verifier to check that every step in the proof follows by the rules of the system. Assume, finally, that you publicize an upper bound for the length of your proof.

The proof can be found nondeterministically in polynomial time, say by a nondeterministic Turing machine. The machine guesses the proof and checks its validity step by step. The computation of the machine can be described in terms of a propositional formula α in 3-conjunctive normal form such that α is satisfiable exactly in case the theorem has a proof whose length does not exceed the given upper bound. Anybody knowing a proof also knows a truth-value assignment showing that α is satisfiable. By Theorem 4.12, you are able to convince a verifier that you know a proof without giving away any information about it except an upper bound on its length. (An upper bound is needed in order to make α a finitary formula.) On the other hand, the verifier is not able to claim and convince anyone else that she knows a proof.

One-way functions are essential in the above considerations. If the verifier is able to open the locked boxes, she learns everything.

The probability of P cheating decreases very rapidly with respect to the number of rounds in the protocols discussed above. Arbitrarily high security cannot be obtained if the number of rounds is bounded. However, the technique can be modified in such a way that this becomes possible. Even *non-interactive* zero-knowledge proofs are possible, such as the one in the following scenario. P and V together generate a long random sequence of bits. After that P leaves for a trip around the world. Whenever he discovers a theorem, he sends a postcard to V, proving his new theorem in a zero-knowledge manner. Since P has no predictable address, interactive protocols are not applicable in this case.

Chapter 5

The Secret Number

5.1 Compressibility of information

How much can a given piece of information be compressed? This is a matter of fundamental scientific importance and is also very relevant for undecidability considerations. In this last chapter we consider the fundamental problem of compressibility of information. Our considerations lead to a mysterious number Ω ("the secret number", "the magic number", "the number of wisdom", "the number that can be known of but not known", to mention only a few possible descriptions) that encodes our cornerstones of undecidability very compactly.

The number Ω is between 0 and 1. It encodes, for instance, the Post Correspondence Problem in the following sense. Suppose we are given an instance PCP. Then, if we know a suitable initial segment of the binary representation of Ω, we are able to compute whether or not PCP possesses a solution. If we restrict our attention to instances of a reasonable size, the knowledge of the first 10 000 bits of Ω is already more than enough. Roughly, if we know the first 10 000 bits of Ω, we are able to solve the halting problems of Turing machines describable in less than 10 000 bits. This surely includes the Turing machines looking for counterexamples of the most famous conjectures in mathematics, such as Fermat's last problem and Riemann's hypothesis. A step further can even be taken. Consider a formal system F of mathematics with axioms and rules of inference. For any well-formed formula α, design a Turing machine $TM(F, \alpha)$ that checks through all the proofs of F and halts if it finds one for α. A similar Turing machine $TM(F, \sim \alpha)$ is designed for the negation of α. Again, if we know a sufficiently long initial segment of Ω then we can decide whether α is provable, refutable or independent. The required length depends on F and α — in all reasonable cases the first 10 000 bits of Ω will suffice. This surely justifies the attributes attached above to Ω.

We use bits as fundamental units of information. A bit indicates a choice between two possibilities, and any piece of information can be encoded as a sequence of bits. Certainly in many cases there are redundant bits in the sequence. In other words, the same sequence can be described in some other way that gives rise to a shorter sequence. This is *per se* the case if the bit sequence is the result of encoding some text in a natural language. Natural languages abound with redundancies that have been estimated numerically for various languages and various types of text. The existence of redundancies is obvious because noise in the information channel may distort or delete some letters, and still nothing is lost in the piece of information. Also, in classical cryptography cryptanalytic attacks are often based on redundancy properties (unbalanced frequency of individual letters, pairs of letters, etc.) of natural languages.

We now investigate the possibility of compressing information, making sequences of bits shorter. Our first goal is to formalize the notion of *compressibility*. The following example contains some additional aspects to be considered.

Example 5.1. Consider the following two sequences of bits, both of length 16:

$$1\ 0\ 1\ 0\ 1\ 0\ 1\ 0\ 1\ 0\ 1\ 0\ 1\ 0\ 1\ 0$$
$$0\ 1\ 1\ 0\ 0\ 0\ 0\ 1\ 1\ 0\ 1\ 1\ 1\ 1\ 1\ 1$$

The first follows an obvious pattern: 10 written eight times. No such uniform pattern is visible in the second sequence. In fact, the second sequence was generated by coin tosses. Tossing the coin sixteen times can produce each of the 2^{16} binary sequences of length 16, and each one of them, including the two mentioned above, has exactly the same probability. Still, it is more difficult to believe of the first than of the second that it results from coin tosses.

The pattern can be used to compress the first sequence. However, even under some favourable notational conventions, "8 times 10" does not necessarily compress it to less than 16 bits. The matter is entirely different if we are dealing with longer sequences of bits. "1 048 576 times 10" certainly describes the sequence of repetitions more compactly than the sequence of 2 097 152 bits 1010...10. Further advantage can be taken of the fact that $1\,048\,576 = 2^{20}$. In general, the most compact way of describing a short sequence is just to write down the sequence.

There may be other ways to compress information than detected patterns. There is no pattern visible in tables of trigonometric functions. Even tables of a modest size give rise to a rather long sequence of bits if everything is expressed as a single sequence. However, a much more compact way to convey the same information is to provide instructions for calculating the tables from the underlying trigonometric formulas. Such a description is brief and, moreover, can be used to generate tables of any size.

Usually no such compression method can be devised for tables presenting empirical or historical facts. For instance, there are books presenting the results of the gold medal winners in each event in each of the Olympic Games since 1896. As regards such information, the amount for compression is negligible, especially if attention is restricted to the least significant digits. Since the results tend to improve, there are regularities in the most significant digits, even to the extent that predictions can be made for the next Olympic Games. In general, as regards empirical data, compression can be linked with *inductive inference* and inductive reasoning as it is employed in science: observations presented as sequences of bits are to be explained and new ones are to be predicted by theories. For our purposes in this chapter, it is useful to view a theory as a computer program to reproduce the observations actually made. The scientist searches for minimal programs, and then the amount of compression can be measured by comparing the size of the program with the size of the data. This also leads to a plausible definition concerning what it means to state that the data are *random*: no compression is possible. In other words, the most concise way of representing the data is just to list them.

Randomness will be a central theme in this chapter. We want to mention one important aspect at this point. There are various statistical tests for disclosing deviations from randomness. Here we consider potentially infinite sequences of bits or digits, such as the decimal expansion of π. A sequence is *normal* iff each bit or digit, as well as each block of any length of bits or digits, occurs with equal asymptotic frequency. Clearly, if a sequence does not qualify as normal, then it is not intuitively viewed as a random sequence. On the other hand, normality does not guarantee randomness, no matter whether we view randomness intuitively or in the sense of incompressibility. For instance, *Champernowne's number*

$$0.12345678910111213141516171819 2021\ldots,$$

obtained by writing all integers one after the other, is normal. Intuitively it is nonrandom. For any i, the ith digit is easily calculated from i and, consequently, long initial segments can be compressed arbitrarily because the formula for the ith digit is independent of the length of the segment. A similar compression is possible for the decimal expansion of the number π because the whole expansion can be generated by a fixed program. Although no proof has been given so far, it is generally conjectured that the expansion of π is normal in the sense described above. The observed statistical data also support this conjecture.

The identification of randomness and incompressibility is also feasible because of the following reason. A gambler who knows the rule governing π or Champernowne's number wins an infinite gain against a gambling casino if the latter produces "random" numbers according to one of the two sequences. Clearly, an intuitive precondition for a sequence being random is that no gambling strategy can produce an infinite gain against that sequence. The knowledge of an initial segment in a random sequence, as opposed to complete ignorance of it, gives no advantage for bets concerning the continuation.

□

The facts presented in Example 5.1 can be viewed as an introduction to the subsequent formal discussion. Example 5.1 shows that some central issues in science, such as inductive reasoning and randomness, are just different aspects of compressibility. In order to formalize the idea that "no program shorter than the sequence can produce the sequence", we have to formalize the notion of a program. For this purpose the following abstract notion of a computer, due to Chaitin, is very suitable. An intuitive picture of the abstract notion will be given in Example 5.2.

A language L, finite or infinite, is termed *prefix-free* iff no word in L is a prefix of another word in L. For instance, the language $\{a^i b | i = 0, 1, 2, \ldots\}$ is prefix-free. The only prefix-free language containing the empty word λ is the language $\{\lambda\}$.

Words in a prefix-free language L can be used to encode letters from another alphabet. For instance, the subset $\{a^i b | 0 \leq i \leq 5\}$ of the above language encodes the letters of the alphabet $V_6 = \{a_0, a_1, a_2, a_3, a_4, a_5\}$ in a natural fashion. Moreover, this encoding has the property of unique decodability: a word over the alphabet $\{a, b\}$ can be decoded in at most one way as a word over V_6. For instance, the word $bba^2 ba^4 b$ can be decoded only as $a_0 a_0 a_2 a_4$, whereas ba can be decoded in no way. The property of unique decodability is always satisfied when words in a prefix-free language L are used to encode letters of an alphabet V. For, assume that the encodings of two different words $a_1 \ldots a_i$ and $b_1 \ldots b_j$, where the as and bs are letters, coincide. We may assume that $a_1 \neq b_1$ because, otherwise, we can divide by the encoding of a_1 from the left. But then the encodings of the two words coincide only if the encoding of a_1 is a prefix of the encoding of b_1, or vice versa, which contradicts the prefix-freeness of L. The argument also shows that decoding from left to right is instantaneous: the remainder of the word does not affect the decoding.

The inequality presented in the next theorem, customarily referred to as the *Kraft inequality*, is important in our discussions concerning probabilities. Although we mostly consider binary alphabets in what follows, the theorem is presented for alphabets with $p \geq 2$ letters.

For a word w over the alphabet $V_p = \{a_1, \ldots, a_p\}$, we define the *frequency indicator* of w, in symbols $FI(w)$, by

$$FI(w) = p^{-|w|} .$$

For a language L over V_p, we define the frequency indicator $FI(L)$ to be the sum of the frequency indicators of the words in L. Thus, $FI(L)$ is a rational number if L is finite. If L is infinite, $FI(L)$ is either a real number or ∞ (depending on whether or not the series converges).

Theorem 5.1. Every prefix-free language L satisfies $FI(L) \leq 1$.

Proof. If $FI(L) > 1$ holds for an infinite language L, then $FI(L_1) > 1$ also holds for a finite subset L_1 of L. Moreover, every subset of a prefix-free language is prefix-free. Consequently, it suffices to establish the claim for a given finite language L.

We claim that, for all $i \geq 1$,

$$(*) \qquad FI(L^i) = (FI(L))^i .$$

$(*)$ is established inductively, the basis $i = 1$ being obvious. Assuming that $(*)$ holds, consider L^{i+1}. Clearly, L^{i+1} is prefix-free and all of its words can be represented uniquely in the form xy, where $x \in L$ and $y \in L^i$. Assume that $L = \{x_1, \ldots, x_r\}$ and denote

$$FI(x_j) = p^{-|x_j|} = t_j, \quad 1 \leq j \leq r .$$

Then
$$FI(L^{i+1}) = \sum_{j=1}^{r} t_j FI(L^i) = FI(L^i) \sum_{j=1}^{r} t_j$$
$$= (FI(L))^i FI(L) = (FI(L))^{i+1}.$$

Here the inductive hypothesis, the unique product representation mentioned above, as well as the obvious equation $FI(xy) = FI(x)FI(y)$, have been used.

Let m be the length of the shortest and M the length of the longest word in L. (Possibly $m = M$.) Consequently, every word in L^i is of length $\geq mi$ and $\leq Mi$. This means that there are at most $(M - m)i + 1$ different lengths possible among the words of L^i. On the other hand, $FI(L_1) \leq 1$ clearly holds for any language L_1 such that all words in L_1 are of the same length. These observations give us the estimate
$$FI(L^i) \leq (M - m)i + 1$$
and hence, by $(*)$,

$(**)$ $\qquad (FI(L))^i \leq (M - m)i + 1$, for all $i \geq 1$.

Since $M - m$ is a constant independent of i, $FI(L) > 1$ would contradict $(**)$ for values of i large enough.

□

The inequality of Theorem 5.1 need not be strict. The language $L = \{a^i b | i = 0, 1, 2, \ldots\}$ considered above satisfies
$$FI(L) = \frac{1}{2} + \frac{1}{4} + \frac{1}{8} + \ldots = 1.$$

The equation also holds for some finite languages, for instance,
$$FI(\{a, b\}) = \frac{1}{2} + \frac{1}{2} = FI(\{a, ba, bb\}) = \frac{1}{2} + \frac{1}{4} + \frac{1}{4}$$
$$= FI(\{aa, ab, ba, bb\}) = \frac{1}{4} + \frac{1}{4} + \frac{1}{4} + \frac{1}{4} = 1.$$

All these languages are maximal with respect to prefix-freeness: if a new word over $\{a, b\}$ is added to the language then the prefix-freeness is lost. This follows either by direct argument or by Theorem 5.1. Observe also that the converse of Theorem 5.1 fails: the inequality $FI(L) \leq 1$ does not imply that L is prefix-free.

We are now ready to introduce the formal notion of a computer. We consider the binary alphabet $V = \{0, 1\}$. (Thus, FI will later be defined for $p = 2$.) For partial functions $f : V^* \times V^* \to V^*$, we also consider "projections" $f_y : V^* \to V^*$, $y \in V^*$, defined by
$$f_y(x) = f(x, y).$$

By definition, a *computer* is a partial recursive function $C : V^* \times V^* \to V^*$ such that, for all $v \in V^*$, the domain of C_v is prefix-free.

Hence, the basic requirement is that whenever $C(u, v)$ is defined (converges) and u is a prefix of u' with $u \neq u'$, then $C(u', v)$ is not defined (diverges).

Example 5.2. We now explain the above definition in terms of a concrete machine model that can be viewed as a modification of the Turing machine. Briefly, u is the *program* and v is the *input*. Here the "program" is understood as a description of the computing strategy in the same sense as a universal Turing machine is given a description of an individual Turing machine, which it is supposed to simulate. Exactly as in case of ordinary Turing machines, the step by step moves of our computer follow a pregiven finite table that completely determines the computation for the argument (u, v).

The computer C has two tapes, a *program tape* and a *work tape*. The program tape is finite. Its leftmost square contains a blank, and each of the remaining squares contains either a 0 or a 1. It is a read-only tape, and the reading head can move only to the right. At the start of the computation, the program u occupies the whole program tape except for the leftmost blank square which is scanned by the reading head.

The work tape is potentially infinite in both directions. Each of its squares contains a blank, 0 or 1. At the beginning all squares are blank except that the input v is written on consecutive squares, and the read–write head scans the leftmost of them. The read–write head can move in both directions. The computer has finitely many internal *states*, among which are a specific *initial* and a specific *halt* state. At the start of the computation, the computer is in the initial state, and no further action is possible in the halt state. (A minor additional detail is that u and/or v may equal the empty word λ.)

The behaviour is defined similarly as for ordinary Turing machines. The triple consisting of the current state and the currently scanned symbols from the program and work tapes (there are three possibilities for both of the latter) determine each of the following: (i) the next state, (ii) the symbol written on the work tape (three possibilities), (iii) the move of the reading head (stay or move one square to the right), (iv) the move of the read–write head (stay or move one square to the right or move one square to the left). Thus, each particular computer can be defined by specifying its behaviour by a finite table.

The computation of a computer C, started with a program u and input v, is a success if C enters the halt state when the reading head is scanning the rightmost square of the program tape. In this case the output value $C(u, v)$ is read from the work tape, starting from the square scanned by the read–write head and extending to the first blank square. Thus, although called briefly "work tape", the work tape also acts as an *input* and an *output tape*. If the computation is a failure, that is, if the halt configuration described above is not reached, then $C(u, v)$ is not defined. The computation of $C(0110, 10) = 0$ is depicted as follows:

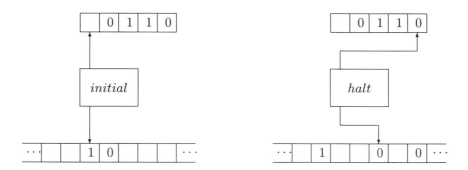

It can now be shown that this concrete machine model computes exactly the partial recursive functions specified in the abstract definition of a computer given before Example 5.2. Indeed, the partial recursive functions computed by the machine model must satisfy the additional condition of prefix-freeness. This follows because the machine is allowed neither to run off the right end of the program tape nor to leave some part of the program unread. Thus, $C(u,v)$ and $C(uu_1,v)$, $u_1 \neq \lambda$, can never both be defined. Observe, however, that $C(u,v)$ and $C(uu_1,v')$ can both be defined.

Conversely, we show how a concrete machine C can simulate an abstract computer C'. The basic idea is that C moves the reading head on the program tape only when it is sure that it should do so.

Given a program u and an input v, C first ignores u and starts generating on its work tape, just as an ordinary Turing machine, the recursively enumerable set $X = \{x | C'(x,v) \text{ is defined}\}$. This is done by dovetailing through computations for all x. Denote by u_1 the prefix of u already read; initially $u_1 = \lambda$. (C keeps u_1 on its work tape.) All the time C keeps checking whether or not u_1 is a prefix of some element of X already generated. If it finds an x such that $u_1 = x$, C goes to the halt state, after first producing the output $C'(u_1,v)$ on its work tape. If C finds an x such that u_1 is a proper prefix of x, then it reads the next symbol a from the program tape and starts comparisons with $u_1 a$.

This works. If $C'(u,v)$ is defined, C will eventually find u. It then compares, step by step, the contents of the program tape with u, and halts with the output $C'(u,v)$ if the program tape contains u. If the program tape contains something else, C does not reach a correct halting configuration. This is also the case if $C'(u,v)$ is not defined. Observe that in this case $C'(u_1,v)$ may be defined for some prefix u_1 of u. Then C finds u_1 and goes to the halt state but the halting configuration is not the correct one: a portion of the program tape remains unread. □

We now continue to develop the abstract notions. By definition, a computer U is *universal* iff, for every computer C, there is a *simulation constant* $\text{sim}(C)$ such that, whenever $C(u,v)$ is defined, there is a program u' (that is, a word over the binary alphabet $V = \{0, 1\}$) such that

$$U(u', v) = C(u, v) \text{ and } |u'| \leq |u| + \text{sim}(C).$$

Theorem 5.2. There is a universal computer.

Proof. Consider an enumeration of all computers, that is, of all tables defining a computer: $C_0, C_1, \ldots$. Clearly, the partial function $F : N \times V^* \times V^* \to V^*$ defined by

$$F(i, u, v) = C_i(u, v), \quad i \in N; \quad u, v \in V^*,$$

is partial recursive, and so is the partial function $U : V^* \times V^* \to V^*$ defined by

$$U(0^i 1 u, v) = C_i(u, v).$$

Moreover, for each v, the domain of the projection U_v is prefix-free. This follows because all projections of each C_i possess the required property of prefix-freeness. Consequently, U is a universal computer with $\text{sim}(C_i) = i + 1$, for each i.

Thinking of the machine model, after reading i 0s from the program tape, U has on its work tape a description of C_i (part from the input v that it already has at the beginning). When U meets the first 1 on the program tape, it starts to simulate the computer C_i whose description it has on its work tape at that moment. (Alternatively, U can store only i on its work tape and, after seeing 1 on the program tape, compute C_i from i.) □

Universal computers are by no means unique. A different U results, for instance, from a different enumeration of the computers C_i. However, from now on we consider a *fixed universal computer* U and speak of *the universal computer U*.

Consider the following ordering (alphabetical and according to length) of the set of words over $V = \{0, 1\}$:

$$\lambda, 0, 1, 00, 01, 10, 11, 000, 001, \ldots .$$

Thus, words are first ordered according to their length, and then words of same length alphabetically. We use the notation $\min(L)$ for the first word of L. If L is empty, $\min(L)$ is undefined. Given $w \in V^*$, the *canonical program w^** for w is defined by

$$w^* = \min\{u \in V^* | U(u, \lambda) = w\}.$$

Thus, w^* is the first (and consequently also the shortest) program of the universal computer U producing w, when U is started with the empty work tape. The next theorem contains some simple observations about the canonical program.

Theorem 5.3. The function $f : V^* \to V^*$ defined by $f(w) = w^*$ is total. For any w, $U(w^*, \lambda) = w$ and $w^* \neq \lambda$.

Proof. Given w, we consider the computer C such that $C(\lambda, \lambda) = w$. Thus, C just prints w on its originally empty work tape, the whole action being embedded in its internal states. (Clearly, we can also use the abstract definition and conclude that a partial recursive function C with the required properties exists. In what follows we use the abstract and concrete notions interchangeably.) By Theorem 5.2, $U(u', \lambda) = w$, for some u'. Hence, the set $\{u \in V^* | U(u, \lambda) = w\}$ is not empty. Since $f(w)$ is defined for an arbitrary w, we conclude that f is total. The equation $U(w^*, \lambda) = w$ is clear by the definition of w^*, and $w^* \neq \lambda$ follows because the domain of U_λ is prefix-free. (Indeed, assume that $U(\lambda, \lambda) = w$. Consider a word $w_1 \neq w$. There is a w' such that $U(w', \lambda) = w$. Since $w_1 \neq w$ we must have $w' \neq \lambda$. This implies that the domain of U_λ is not prefix-free.) □

We now define the notion of the *complexity* (which could also be called *program-size complexity*) associated to a word $w \in V^*$. By definition, the *complexity* of a word w with respect to a computer C, equals

$$H_C(w) = \min\{|u| \mid u \in V^* \text{ and } C(u, \lambda) = w\}.$$

Again, min is undefined if the set involved is empty. We denote briefly that $H(w) = H_U(w)$.

Thus, $H(w)$ is the length of the minimal program for U to compute w when started with the empty work tape. We can view the ratio between $H(w)$ and $|w|$ as the *compressibility* of w. We want to emphasize that, intuitively, $H(w)$ should be understood as the *information-theoretic* (or program-size) complexity of w, as opposed to the computational complexity of w. This is also why we have chosen the notation H, standard in information theory in entropy considerations. $H(w)$ could also be referred to as the *algorithmic entropy*.

We next define the *conditional complexity* of w, given $t \in V^*$.

By definition, for a computer C, the *conditional complexity* of a word $w \in V^*$ with respect to a word $t \in V^*$ equals

$$H_C(w/t) = \min\{|u| \mid u \in V^* \text{ and } C(u, t^*) = w\}.$$

We denote briefly $H_U(w/t) = H(w/t)$, and speak of the *conditional complexity* of w with respect to t. It is immediate by Theorem 5.3 that $H(w)$ and $H(w/t)$ are defined for all w and t.

The complexities defined depend on the computer C. This is also true as regards $H(w)$ and $H(w/t)$, because they depend on our chosen universal computer. The next theorem asserts that words also possess an inherent complexity, independent of the computer. This holds both for plain and conditional complexity. Although easy to prove, this result is of fundamental importance for the whole theory. The result is often referred to as the *Invariance Theorem*. It says that the universal computer is asymptotically optimal. It does not say much about the complexity of an individual word w, because the constant involved may be huge with respect to the length of w.

Theorem 5.4. For every computer C, there is a constant A_C such that

$$H(w) \leq H_C(w) + A_C \text{ and } H(w/t) \leq H_C(w/t) + A_C,$$

for all w and t.

Proof. The simulation constant $\text{sim}(C)$ satisfies the requirements for A_C and, hence, the theorem follows. Observe that $H_C(w)$ and $H_C(w/t)$ might be undefined. We may either agree that the theorem concerns defined values only, or consider the undefined values as ∞.
□

Recall that we fixed the universal computer U in a rather arbitrary, and certainly in a very nonspecific, way. According to Theorem 5.4, this does not matter. For any two universal computers U and U', there is a constant $A_{U,U'}$ such that

$$(*) \qquad |H_U(w) - H_{U'}(w)| \leq A_{U,U'},$$

for all w. The same result also holds for the absolute value of the difference of the conditional complexities. Observe, however, that an analogous result does not hold for arbitrary computers C and C'.

The equation $H(w) = |w^*|$ is an immediate consequence of the definition of w^* and Theorem 5.3.

Sometimes it is convenient to use the 0-notation for the *order of magnitude*, defined for functions on real numbers. By definition, $f(x) = 0(g(x))$ iff there are positive constants a and x_0 such that

$$|f(x)| \leq a|g(x)|$$

holds for all $x \geq x_0$. Thus, the above inequality $(*)$ can be written as

$$H_U(w) = H_{U'}(w) + 0(1).$$

Let $\varphi(x, y)$ be a *pairing function*, that is, a recursive bijection of $V^* \times V^*$ on to V^*. (Such a pairing function is obtained, for instance, from our ordering of V^* using a pairing function for nonnegative integers, see Section 1.1.) Given a computer C, we define

$$H_C(w, t) = H_C(\varphi(w, t))$$

and again $H(w, t) = H_U(w, t)$. Intuitively, $H(w, t)$ is the length of the shortest program that outputs w and t in a way that tells them apart. The next theorem is an exercise concerning symmetry.

Theorem 5.5. $H(w, t) = H(t, w) + 0(1)$.

Proof. The idea is to use the "inverse components" of φ to carry out the commutation. Thus, let $\varphi_i : V^* \to V^*$, $i = 1, 2$, be functions such that

$$\varphi(\varphi_1(w), \varphi_2(w)) = w$$

for all w. Consider the computer C, defined by the condition

$$C(x, \lambda) = \varphi(\varphi_2(U(x, \lambda)), \varphi_1(U(x, \lambda))).$$

We now use the fact that, for all w,

$$H(w) \leq H_C(w) + \mathrm{sim}(C).$$

Consequently, we compute

$$H(w, t) = H(\varphi(w, t)) \leq H_C(\varphi(w, t)) + \mathrm{sim}(C)$$
$$= \min\{|x| \mid x \in V^* \text{ and } \varphi(\varphi_2(U(x, \lambda)), \varphi_1(U(x, \lambda))) = \varphi(w, t)\} + \mathrm{sim}(C)$$
$$= \min(|x| \mid x \in V^* \text{ and } \varphi_2(U(x, \lambda)) = w \text{ and } \varphi_1(U(x, \lambda)) = t\} + \mathrm{sim}(C)$$
$$= H(\varphi(t, w)) + \mathrm{sim}(C) = H(t, w) + \mathrm{sim}(C).$$

Here also the definition of H_C has been used. Observe also that the mapping $U(x, \lambda) = U_\lambda(x)$ is surjective: the universal computer is capable of producing any word starting with the empty word as its input. □

In fact, the above proof shows that if $g : V^* \to V^*$ is a recursive bijection, then

$$H(w) = H(g(w)) + 0(1).$$

To see this, it suffices to consider the computer C defined by the condition

$$C(x, \lambda) = g(U(x, \lambda)).$$

The above proof uses the recursive permutation

$$g(x) = \varphi(\varphi_2(x), \varphi_1(x)).$$

The next theorem presents further results concerning the interrelation between $H(w)$, $H(w/t)$ and $H(w, t)$. In (ii) the notation $\overline{H(w)}$ means the word over V whose ordinal number in our ordering of V^* equals $H(w)$.

Theorem 5.6

(i) $H(w/w) = 0(1)$.
(ii) $H(\overline{H(w)}/w) = 0(1)$.
(iii) $H(w) \leq H(w, t) + 0(1)$.
(iv) $H(w/t) \leq H(w) + 0(1)$.
(v) $H(w, t) \leq H(w) + H(t/w) + 0(1)$.
(vi) $H(w, t) \leq H(w) + H(t) + 0(1)$.

Proof. Consider (i). Since $H(w/w)$ is obviously always nonnegative, it suffices to show that there is a constant A such that $H(w/w) \leq A$, for all w. Define the computer C by the condition

$$C(\lambda, v) = U(v, \lambda), \text{ for all } v \in V^*.$$

Consequently,
$$C(\lambda, w^*) = U(w^*, \lambda) = w$$

and hence,
$$H_C(w/w) = \min\{|u| \mid u \in V^* \text{ and } C(u, w^*) = w\} = 0.$$

By Theorem 5.4,
$$H(w/w) \leq H_C(w/w) + \text{sim}(C) = \text{sim}(C) = A.$$

For (ii), define the computer C by the following condition. Whenever $U(u, \lambda)$ is defined,
$$C(\lambda, u) = \overline{|u|}.$$

Since $U(w^*, \lambda) = w$, we obtain
$$C(\lambda, w^*) = \overline{|w^*|} = \overline{H(w)}.$$

Consequently,
$$H(\overline{H(w)}/w) \leq H_C(\overline{H(w)}/w) + \text{sim}(C)$$
$$= \min\{|u| \mid u \in V^* \text{ and } C(u, w^*) = \overline{H(w)}\} + \text{sim}(C)$$
$$= 0 + \text{sim}(C) = \text{sim}(C),$$

from which (ii) follows.

For (iii) and (iv), we use similarly the computers C and C' defined by the conditions
$$\begin{aligned} C(u, \lambda) &= \varphi_1(U(u, \lambda)) \text{ and} \\ C'(w, t) &= U(w, \lambda). \end{aligned}$$

Clearly, (vi) is an immediate consequence of (iv) and (v). Hence, it suffices to establish (v). This will be the most interesting construction used in Theorem 5.6.

We *claim* that there is a computer C satisfying the following property. Whenever

(∗) $$U(u, w^*) = t \text{ and } |u| = H(t/w),$$

that is, whenever u is a minimal-size program for U to compute t from w^*, then

(∗∗) $$C(w^* u, \lambda) = \varphi(w, t).$$

Compressibility of information 173

Let us see how (v) follows from this claim. Indeed, by definition, $H_C(w,t)$ is the length of the shortest program for C to compute $\varphi(w,t)$ from the empty word. Since, by (**), w^*u is such a program, we obtain

$$H_C(w,t) \leq |w^*u| = |w^*| + |u| = H(w) + H(t/w) .$$

From this (v) follows because, always, $H(w,t) \leq H_C(w,t) + \mathrm{sim}(C)$.

It remains to verify the claim that there is such a computer C. We follow the abstract definition. Let $C(x,y)$ be the following partial recursive function. For $y \neq \lambda$, $C(x,y)$ is undefined. Let Y be the domain of U_λ, that is,

$$Y = \{u \in V^* | U(u,\lambda) \text{ is defined}\} .$$

The following effective procedure is now used to compute the value $C(x,\lambda)$. Elements of the recursively enumerable set Y are generated until, if ever, a prefix v of x is found. Then we write x in the form $x = vu$ and simulate the computations $U(u,v)$ and $U(v,\lambda)$. If both of these computations halt, we output

$$C(x,\lambda) = \varphi(U(v,\lambda), U(u,v)) .$$

Obviously, C is partial recursive. We show that C satisfies the required condition (**), whenever u, w and t satisfy (*). Denote $x = w^*u$, and consider our algorithm for computing $C(x,\lambda)$. Since $U(w^*,\lambda) = w$, we conclude that w^* is in Y and, consequently, will eventually be found as a prefix v of x, yielding the factorization $x = vu = w^*u$. Moreover, it is not possible that another prefix v' of x would be found in this fashion, because then one of the words v and v' would be a prefix of the other, where both of the words v and v' are in Y. However, this would contradict our basic assumption concerning computers: the domain of U_λ is prefix-free.

Once the unique factorization $x = vu$ with $v = w^*$ has been found, both of the computations $U(u,v)$ and $U(v,\lambda)$ halt because

$$U(u,v) = U(u,w^*) = t \text{ and } U(v,\lambda) = U(w^*,\lambda) = w .$$

Now we see that the output satisfies

$$C(x,\lambda) = \varphi(U(v,\lambda), U(u,v)) = \varphi(w,t) ,$$

as it should according to (**).

We still have to show that $C(x,y)$ is a computer, that is, the domain of C_y is prefix-free, for all y. This is clear for $y \neq \lambda$. Consider the domain of C_λ. Assume that x is a prefix of y and that $C(x,\lambda)$ and $C(y,\lambda)$ are both defined. This implies that we obtain the decompositions

$$x = v_x u_x \;,\; y = v_y u_y$$

and, moreover, each of the values

$$U(u_x, v_x) \,,\; U(v_x, \lambda) \,,\; U(u_y, v_y) \,,\; U(v_y, \lambda)$$

is defined. Since x is a prefix of y, one of the words v_x and v_y is a prefix of the other. But this is possible only if $v_x = v_y$, because the domain of U_λ is prefix-free. We now conclude that u_x is a prefix of u_y. Because both u_x and u_y belong to the prefix-free domain of U_{v_x} ($= U_{v_y}$), we make the further conclusion that $u_x = u_y$ and, consequently, $x = y$. This shows that the domain of C_λ is prefix-free. We have completed the proof of (v).

□

We introduce a further notion. The amount by which the conditional complexity $H(w/t)$ is less than the (unconditional) complexity $H(w)$ can be viewed to indicate how much t contains information about w. The notion is again first defined for arbitrary computers and then for the universal computer.

By definition, the *algorithmic information in t about w* with respect to the computer C equals

$$I_C(t:w) = H_C(w) - H_C(w/t),$$
$$I(t:w) = I_U(t:w).$$

One would expect that the information contained in w about w is approximately the same as the information contained in w, and that the information contained in λ about w, and vice versa, amounts to nothing. Also, the other formal statements presented in the next theorem are plausible from the intuitive point of view.

Theorem 5.7

(i) $I(t:w) \geq 0(1)$.
(ii) $I(t:w) \leq H(t) + H(w) - H(t,w) + 0(1)$.
(iii) $I(w:w) = H(w) + 0(1)$.
(iv) $I(w:\lambda) = 0(1)$.
(v) $I(\lambda:w) = 0(1)$.

Proof. Assertions (i)–(iii) follow by Theorem 5.6, from assertations (iv), (v) and (i), respectively. To prove (iv), we observe the already established (i) and write according to (ii):

$$I(w:\lambda) \leq H(w) + H(\lambda) - H(w,\lambda) + 0(1)$$
$$= H(w) - H(w,\lambda) + 0(1) = 0(1).$$

Here the last equation is established by defining the computer

$$C(u,\lambda) = \varphi_1(U(u,\lambda))$$

and observing that $H(w) \leq H(w,\lambda) + \text{sim}(C)$. The proof of (v) is similar.

□

Compressibility of information 175

The notion of complexity discussed in this chapter could be referred to as the *Chaitin complexity*; the basic concepts and the notation, as well as the definition of the computer C, are from [Cha 1]. This notion of complexity falls within the framework of *descriptional* or *information-theoretic* complexity, as opposed to *computational* complexity. In the former, one is interested in the *program size*: only the length of the shortest program for a specific task matters, not how difficult the computation will be according to the program. For the latter type of complexity, computational complexity, the *amount of resources* such as time needed in the computation is essential.

Example 5.3. Many problems concerning regular languages, although decidable, are known to be difficult from the point of view of computational complexity. For instance, it has been shown that for some such problems no polynomial space bound exists. We use this fact to show that, sometimes, the descriptional complexity is "as low as it can be", whereas the computational complexity is high.

Consider regular expressions over the alphabet $\{a, b\}$. Thus, we use Boolean operations, catenation and star to the "atoms" a, b, ϕ. Let α_i, $i = 0, 1, \ldots$, be an ordering of such regular expressions. For instance, we may order the regular expressions first according to length and then alphabetically to obtain the order α_i. Let us call an index i *saturated* iff the regular language denoted by α_i equals $\{a, b\}^*$, that is, its complement is empty. A real number

$$r = .a_0 a_1 a_2 \ldots$$

in binary notation is now defined by the condition: $a_i = 1$ iff i is saturated.

Clearly, $r > 0$ because some indices i are saturated. On the other hand, in our ordering, which takes the length into account, the first saturated index appears quite late, because at the beginning only languages with a nonempty complement appear. We also consider the language consisting of all prefixes of the binary expansion of r:

$$L_r = \{w \in \{0,1\}^+ | w = a_0 a_1 \ldots a_i, \text{ for some } i \geq 0\}.$$

It is obvious that there is a constant A such that the descriptional complexity of words in L_r is bounded from above by A, provided the length of the word is given. This follows because the algorithm for computing bits of the expansion of r, that is, for testing the emptiness of certain regular languages can be carried out by a fixed computer. Such a computer can be one of the computers C defined above. Then C starts with an empty program tape and with the index i written in binary notation on its work tape, and halts with the output $a_0 a_1 \ldots a_i$. Since the binary expansion of i contains, roughly, $\log i$ bits we obtain the result $H(w) \leq \log |w| + A$ for all w in L_r. The same result is also obtained if r is the decimal or binary expansion of π. This can be viewed as the lowest possible descriptional complexity: the same algorithm produces arbitrarily long prefixes of an infinite sequence. This is a property shared by all computable (recursive) sequences. The estimate $H(w) \leq \log |w| + A$ can be further improved by replacing $|w| = i+1$ with its descriptional complexity. This yields the estimate $H(w) \leq H(i+1) + A$.

On the other hand, the computational complexity of r (which can be viewed as the computational complexity of the membership problem in L_r) is very high. Indeed, in order to decide the membership of a word of length i, one has to settle the emptiness of the complement for all of the i regular languages involved. Although both r and π are of essentially the same low descriptional complexity, the computational complexity of r is essentially higher than that of π.

It is natural to ask whether there are reverse examples, that is, cases where the computational complexity is low but the descriptional complexity is high. This is very much a matter of how the setup is defined. No examples as obvious as those given above can be given.

□

The descriptional complexity of a word, or the amount of information in a word, can be identified as the size of the smallest program that produces the word when started with a blank memory. In general, according to different models, the amount of information is invariant up to a minor additive term. Descriptional complexity is nowadays usually referred to as the *Kolmogorov complexity*. With the early developments from the 1960s in mind, *Solomonoff–Chaitin–Kolmogorov complexity* is perhaps the most appropriate term for descriptional complexity. (See [Li Vi] for a historical account concerning these matters.) Our notion of Chaitin complexity could also be called *self-delimiting Kolmogorov complexity*. We now discuss this difference in more detail, and also explain why Chaitin complexity is more appropriate for our purposes.

In general, the *Kolmogorov complexity* of a word w with respect to a description method M, in symbols $K_M(w)$, is the length of the shortest program p such that M with the program p produces w. To obtain a formal definition, we identify "description methods" with partial recursive functions.

Consider again the binary alphabet $V = \{0, 1\}$, and let $f : V^* \times V^* \to V^*$ be partial recursive. Then the *Kolmogorov complexity* of a word w in V^* with respect to f is defined by

$$K_f(w) = \min\{|p| \mid f(p, \lambda) = w\}.$$

The min-operator is undefined when applied to the empty set. Similarly, the *conditional Kolmogorov complexity* of w, given t, is defined by

$$K_f(w/t) = \min\{|p| \mid f(p, t) = w\}.$$

Let us immediately compare this definition with our previous definition of (Chaitin) complexity and conditional complexity. The essential difference is that the definition of the Kolmogorov complexity does not require the condition concerning prefix-freeness: it is not required that the domain of

$$f_t(p) = f(p, t)$$

is prefix-free for all t. We will comment on this further in the following. Another difference is that t rather than t^* appears as the second argument place of f in the definition of the conditional Kolmogorov complexity. (One consequence of this difference is that $K(w/\lambda) = K(w)$, whereas in general $H(w/\lambda) \neq H(w)$.)

The *Invariance Theorem* holds for the Kolmogorov complexity as well, the proof being essentially the same as for Theorem 5.4.

Theorem 5.8. There is a partial recursive function f_U (a "universal" function) with the following property. For every partial recursive function f, there is a constant A_f such that

$$K_{f_U}(w) \leq K_f(w) + A_f \quad \text{and} \quad K_{f_U}(w/t) \leq K_f(w/t) + A_f,$$

for all w and t.

As before, we fix a partial recursive function f_U satisfying Theorem 5.8 and use the simple notation K instead of K_{f_U}. We can also define, as before,

$$K(w, t) = K(\varphi(w, t)),$$

where φ is a pairing function. But a remarkable difference is that now we do *not* obtain the estimate $K(w, t) \leq K(w) + K(t) + 0(1)$, analogous to Theorem 5.6 (vi). Let us look into this in more detail.

The construction for Theorem 5.6 (v), used to obtain Theorem 5.6 (vi), does not work for K instead of H. If arbitrary partial recursive functions rather than computers (satisfying the prefix-freeness condition) are used, then the decomposition $x = vu$ is not clear *a priori* but additional information is needed to differentiate v and u. This amounts to information concerning the length of either w or t. The length of w being logarithmic in terms of w viewed as a binary number, the estimate corresponding to Theorem 5.6 (vi) reads

$$K(w, t) \leq K(w) + K(t) + 0\left(\log(\min(K(w), K(t)))\right).$$

It can be shown that the logarithmic fudge term is necessary.

We have already pointed out that the Chaitin complexity could be called *self-delimiting Kolmogorov complexity*. Here "self-delimiting" means that the total length of a program must be given within the program itself. In our machine model (the computer C) this is taken care of by the requirement that the computer never runs off the end of the program tape or ignores a part of the program. This leads to prefix-freeness: no program for a successful computation is a prefix of another.

Programs are self-delimiting if constructs are provided for beginning and ending a program. This is easily accomplished if end markers are available. The situation is trickier if the whole program is, as in our case, just a sequence of bits. The program might actually describe a formal parametrized procedure, as well as list the values of the parameters. One has to be able to differentiate between all these items in the single-bit sequence that constitutes the program. How this requirement (of self-limitedness) can be realized will be illustrated in the next example.

Example 5.4. Given a word w over the binary alphabet, we define the *self-delimiting presentation* $SD(w)$ of w. The idea is to write the length of w written in binary notation in front of w, that is, to consider the word $|w|w$, where $|w|$ is given in binary notation. The following additional trick makes it possible to recognize immediately in a (possibly long) binary sequence, where the length indicator ends and the "proper word" begins. (Although every word over V can be represented in at most one way as the catenation of $|w|$ and w, for some w, an unbounded amount of lookahead is needed to differentiate the border between $|w|$ and w without the additional trick.) In $|w|$ we write a 0 between every two bits and, moreover, add a bit 1 to the end. In this fashion we obtain the word $ML(w)$, the "modified length" of w. Finally, $SD(w) = ML(w)w$. Some illustrations are given in the following table:

w	$|w|$	$ML(w)$	$SD(w)$
01	10	1001	100101
110101	110	101001	101001110101
0^{15}	1111	10101011	$10101011 0^{15}$

Consider now an arbitrary word x over $\{0, 1\}$. To present x in the form $x = ML(w)w$, provided this is at all possible, it suffices to find the first bit 1 in x that occurs in an even-numbered position, counted from left to right.

The longer the word w, the less is the contribution of $ML(w)$ to the length of $SD(w)$. Asymptotically, we have

$$|SD(w)| = |w| + 2\log|w|.$$

Thus, the length increases only by an additive logarithmic term in the transition from a word to its self-delimiting presentation. Such a difference by a logarithmic term is often presented when comparisons are made between the Kolmogorov complexity and the self-delimiting Kolmogorov complexity (Chaitin complexity). A typical example is Theorem 5.6 (vi). Because K is not self-delimited, a logarithmic term is needed, but it is not needed for H because it is already taken care of in the definitions.

□

To summarize, one can distinguish in the study of descriptional complexity the self-delimiting version (Chaitin complexity) and the nonself-delimiting version (Kolmogorov complexity). As regards conditional complexity, there is a further possibility for a different definition, depending on whether a word t itself, or the shortest program t^* for it is considered. (We made the latter choice in our definition of complexity.) Different variations in the basic definitions result in somewhat different theories but it is beyond the scope of this book to investigate this matter in detail.

For our purposes, the given definition of complexity, using computers C and the resulting prefix-freeness, is the most suitable. Our main purpose in this chapter is

Randomness and magic bits 179

the discussion concerning the secret number Ω, a very compact way of encoding the halting problem (or any of the cornerstones of undecidability). In this discussion, probabilities play a central role: we will use the Kraft inequality (Theorem 5.1). The self-delimiting version satisfies the essential requirement for prefix-freeness.

5.2 Randomness and magic bits

We now return to the discussion, started at the beginning of the preceding section, concerning randomness viewed as incompressibility. Before giving the formal definitions, we will begin with some informal considerations. As before, we will talk about words w over the binary alphabet $V = \{0, 1\}$.

A word w is incompressible or random iff the shortest program describing w is roughly of the same length as w. An infinite sequence of bits is random iff this condition holds for all prefixes of the sequence. For (finite) words w, randomness is a matter of degree. Depending on the model, a certain term additional to $|w|$ is needed, and the degree of randomness indicates how close the length of the shortest program for w is to the maximal value. "How random is w?" is the proper question to ask in this case because the additional term may be huge or negligible, depending on the length of w. For infinite sequences, there is a sharp distinction between randomness and nonrandomness, because the additional term will eventually become negligible.

Consider the degree of randomness of a word w with $|w| = i > 20$. Call a word w "fairly random" iff the shortest program for describing w is of length $\geq i-20$. (Here we have an arbitrary model in mind, not necessarily the model with computers C and complexity H.) The following argument shows that almost all words are fairly random. Consider a fixed i. Assume that every word over V of length $< i - 20$ is actually a program describing a word of length i. There are

$$2^1 + 2^2 + \cdots + 2^{i-21} = 2^{i-20} - 2$$

nonempty words of length $< i - 20$. Hence, out of the 2^i words of length i, at most $2^{i-20} - 2$ can be described by a program of length $< i - 20$. This holds for an arbitrary i. The ratio between $2^{i-20} - 2$ and 2^i is less than 10^{-6}. Consequently, less than one in a million among the numbers of any given length is not fairly random. If randomness is taken to mean information-theoretic incompressibility in the sense explained above, almost all numbers are fairly random. A formal counterpart of this result will be presented in Theorem 5.12. If a word w over $\{0, 1\}$ is generated by fair coin tosses, the probability is greater than 0.999 999 that the result will be random to the extent indicated in the definition of fair randomness. This would seem to suggest that it is easy to exhibit a specimen of a long fairly random word. One of the conclusions made in the following will be that it is actually impossible to do so.

Let us elaborate the latter claim. The claim is that, although most words are random, we will never be able to explicitly exhibit a long word that is demonstrably random. The reason is that the axioms and rules of inference in any formal system T can be described in a certain number, say n, of bits. The system T cannot be used to prove the randomness of any word w much longer than n because, otherwise, a contradiction would arise. If $|w| = m$ is much larger than n and the randomness of w could be proven within T, we could construct a computer C to check through the proofs of T, until the correct one is found. The description of C takes roughly n bits and, thus, we obtain a program much shorter than m describing w, a contradiction. To put it very simply, we cannot construct a program p to print a word w with $|w| > |p|$ unprintable by programs q with $|q| < |w|$. Chaitin has expressed the matter by saying that if one has ten pounds of axioms and a twenty-pound theorem, then the theorem cannot be derived from the axioms.

Thus, we cannot know a random number but we can still know of a random number. In particular, we can know of the secret number Ω: it is the probability that the universal computer U halts when it is started with an empty work tape. (Thus U is started with a pair (u, λ), where the program u is arbitrary.) Before going into formal details, we present another possibility for encoding the halting problem.

Example 5.5. We now consider ordinary Turing machines because self-delimitedness is not important here. Let TM_0, TM_1, TM_2, ... be a numbering of all Turing machines, and define a number A by its binary expansion

$$A = .a_0 a_1 a_2 \ldots ,$$

where, for all i, $a_i = 1$ iff TM_i halts with the empty tape. We have already pointed out in Example 5.3 that computable sequences are never random. However, A is clearly noncomputable and, thus, could be random as far as this matter is concerned. But A is not random. A gambler is able to make an infinite profit by using some infinite subclass of Turing machines with a decidable halting problem. A formal argument concerning the compressibility of A can be based on the following observation. Consider any prefix of A of length n. Suppose we know the *number* m of 1s in this prefix. Then we also know the prefix itself, because we can dovetail the computations of the first n Turing machines until we have found m of them that have halted. Eventually, this will happen. Thus, information about the prefix of length n can be compressed to information about n and m.

□

We are now ready for the formal definitions. We consider the binary alphabet $V = \{0, 1\}$. The definition of a computer C and a universal computer U are the same as in Section 5.1. Moreover, as was also done in Section 5.1, a fixed universal computer U will be considered throughout. The "optimal program" t^* for t is defined as before. By definition, *the probability* of a word $w \in V^*$ with respect to a computer C equals

$$P_C(w) = \sum_{\substack{u \in V^* \\ C(u,\lambda) = w}} 2^{-|u|} .$$

The *conditional probability* of w with respect to a word $t \in V^*$ and computer C equals

$$P_C(w/t) = \sum_{\substack{u \in V^* \\ C(u,t^*)=w}} 2^{-|u|}$$

$P_C(w)$, resp. $P_C(w/t)$, is defined to be 0 if no u as required on the right side exists. The *probability* of w and the *conditional probability* of w with respect to t are defined by

$$P(w) = P_U(w) \quad \text{and} \quad P(w/t) = P_U(w/t).$$

Finally, the *halting probability of the universal computer* is defined by

$$\Omega = \sum_{\substack{u \in V^* \\ U(u,\lambda) \text{ converges}}} 2^{-|u|}.$$

Instead of probabilities, we could speak of *algorithmic probabilities* or *information-theoretic probabilities*. The justification for the terminology and the interconnection with the classical probability theory will be discussed below. We first prove some formal results.

Theorem 5.9. *The following inequalities hold for all words w and t over V and for all computers C:*

(i) $0 \leq P_C(w) \leq 1$,
(ii) $0 \leq P_C(w/t) \leq 1$,
(iii) $0 \leq \sum_{x \in V^*} P_C(x) \leq 1$,
(iv) $0 \leq \sum_{x \in V^*} P_C(x/t) \leq 1$.

Proof. Recall the definition of the frequency indicator, FI, given in Section 5.1. Clearly,

$$P_C(w) = FI(L), \quad \text{where} \quad L = \{u \in V^* | C(u,\lambda) = w\}.$$

By the definition of a computer, the domain L' of C_λ is prefix-free. L is a subset of this prefix-free language and, hence, L itself is prefix-free. The inequality $P_C(w) \leq 1$ now follows by Theorem 5.1.

Observe, next, that $\sum_{x \in V^*} P_C(x) = FI(L')$, and hence the upper bound in (iii) follows by Theorem 5.1. To obtain the upper bounds in (ii) and (iv), we use the domain of C_{t^*} in the same way (the domain being prefix-free by the definition of a computer) and its subset determined by w. The lower bounds in (i)–(iv) are obvious because all terms in the series are nonnegative. □

The next theorem presents an interconnection between probability and complexity.

Theorem 5.10. For all w, t and C,
$$P_C(w) \geq 2^{-H_C(w)} \quad \text{and} \quad P_C(w/t) \geq 2^{-H_C(w/t)}.$$

Proof. According to the definition, $P_C(w)$ (resp. $P_C(w/t)$) is a sum of terms, one of which is $2^{-H_C(w)}$ (resp. $2^{-H_C(w/t)}$). The inequalities are also formally valid for undefined values of H_C if we agree that the value is ∞ in this case. □

Theorem 5.11. For all w and t,
$$0 < P(w) < 1 \quad \text{and} \quad 0 < P(w/t) < 1.$$

Proof. The inequalities $0 < P(w)$ and $0 < P(w/t)$ follow by Theorem 5.10 because $H(w)$ and $H(w/t)$ are always defined (see Theorem 5.3). By Theorem 5.9 (iii),
$$\sum_{x \in V^*} P(x) \leq 1,$$
and, as we have already observed that each term in the sum is greater than 0, we must have $P(w) < 1$. The inequality $P(w/t) < 1$ follows similarly by Theorem 5.9 (iv). □

We next establish some upper bounds concerning the cardinalities of certain sets defined in terms of H and P.

Theorem 5.12. For all computers C, words t and integers $m, n \geq 1$, we have

(i) $\text{card}\{w \in V^* | H_C(w) < m\} < 2^m$,
(ii) $\text{card}\{w \in V^* | H_C(w/t) < m\} < 2^m$,
(iii) $\text{card}\{w \in V^* | P_C(w) > \frac{m}{n}\} < \frac{n}{m}$,
(iv) $\text{card}\{w \in V^* | P_C(w/t) > \frac{m}{n}\} < \frac{n}{m}$.

Proof. $H_C(w)$ is the length of the shortest u, if any, such that $C(u, \lambda) = w$. Each u gives rise to at most one w, and there are no more than $2^m - 1$ possible us, since this is the total number of words shorter than m. Hence, (i) follows. The proof for (ii) is similar. Arguing indirectly, we see that if (iii) does not hold, then
$$1 = \frac{n}{m} \cdot \frac{m}{n} \leq \text{card}\{w \in V^* | P_C(w) > \frac{m}{n}\} \cdot \frac{m}{n}$$
$$< \sum_{w \in V^*} P_C(w) \leq 1.$$

Here the last inequality follows by Theorem 5.9 (iii), and the strict inequality holds because we have strictly increased every element in a sum and possibly added new elements. The contradiction $1 < 1$ shows that (iii) holds. Again, the proof for (iv) is similar. □

We want to emphasize at this stage that in our definitions above the term "probability" is only suggestive. No properties of probabilities have been used in the proofs of Theorems 5.9–5.12. However, this suggestive term is very well justified. Intuitively, $P_C(w)$ coincides with the probability for the computer C to produce the output w when started with a program u generated by coin tosses, and an empty work tape. (The length of the program tape is adjusted according to the program u.) An analogous intuitive point of view can be given as regards the other P-notions. A more formal approach would be to introduce a uniform probabilistic structure s on the alphabet V by defining $s(0) = s(1) = \frac{1}{2}$, and to consider the product space, consisting of infinite sequences of letters of V, provided with the product probability. The details of such an approach are of no concern to us in this book.

We now return to the discussion of randomness. We have already pointed out that, for words w, there is no sharp distinction between randomness and nonrandomness. However, one can speak of the *degree of randomness*. For a word w of length i, the degree of randomness of w indicates how close $H(w)$ is to the maximal value $i + H(i)$.

As regards infinite sequences, there is a sharp distinction. We can define explicitly what we mean by a random sequence. Although practically all infinite sequences are random, it is (and will be!) impossible to exhibit one which we can prove is random.

For an infinite sequence

$$B = b_1 b_2 b_3 \ldots$$

of elements of $\{0,1\}$, we let $B_i = b_1 \ldots b_i$ be the prefix of length i of B, for $i = 1, 2, \ldots$. By definition, the sequence B is *random* iff there is a constant A such that, for all i,

$$H(B_i) > i - A .$$

In Section 5.1, we observed that random sequences must possess certain statistical properties such as being normal. Such properties must be present in all reasonably constructed subsequences as well. In Example 5.5, for instance, we found a reasonable subsequence consisting of 1s only — hence the whole sequence is not random. One can formulate the notion of randomness in terms of effectively verifiable statistical tests. What is very pleasing is that this definition of randomness yields exactly the same random sequences as the definition given above, see [Li Vi] and [Cal].

We now discuss the properties of Ω, the halting probability of the universal computer, the "secret number". Taking $C = U$ in Theorem 5.9 (iii), we first obtain $0 \leq \Omega \leq 1$. Theorem 5.11 shows that the first inequality is strict. That also the second inequality is strict is a consequence of the fact that U cannot halt for all programs used in the proof of the Kraft inequality, Theorem 5.1. Thus, we have

$$0 < \Omega < 1 .$$

Let
$$\Omega = .b_1 b_2 b_3 \ldots$$
be the binary expansion of Ω. Ambiguities are avoided by choosing the non-terminating expansion whenever two expansions are possible. (We do not want to exclude the possibility of Ω being rational!) The bits b_i, $i \geq 1$, in the expansion of Ω are referred to as *magic bits*. Thus, Ω is a real number. We also consider the infinite sequence
$$B = b_1 b_2 b_2 \ldots$$
of letters of $\{0, 1\}$, as well as its prefix of length i:
$$B_i = b_1 \ldots b_i, \quad i \geq 1.$$
For all $i \geq 1$, we define the rational number
$$\Omega_i = .b_1 \ldots b_i.$$

We are now ready to establish the remarkable properties of the number Ω hinted at earlier in this chapter. We first prove a result showing that Ω encodes the halting problem in a very compact form. We then establish that B (as defined above from Ω) is random, and proceed to consider the implications for formal axiomatic theories. Any theory is capable of yielding only finitely many bits of B. Thus, we can never know Ω in the sense that we could somehow produce infinitely many bits of B, but we can know of Ω in the sense that we can define it formally. Finally, we show that no formal axiomatic theory can ever determine whether a certain parametrized Diophantine equation has infinitely many solutions, except for finitely many parameter values.

The domain of U_λ, that is, the set
$$DOM(U_\lambda) = \{u \in V^* | U(u, \lambda) \text{ is defined }\}$$
is recursively enumerable. We consider some fixed enumeration of it, where repetitions do not occur. Such an enumeration is obtained by dovetailing. Thus, we obtain a total recursive injection g of the set of positive integers into V^*. We define, for $n \geq 1$,
$$\omega_n = \sum_{j=1}^{n} 2^{-|g(j)|}.$$
It is obvious that the sequence ω_n, $n = 1, 2, \ldots$, is strictly monotonically increasing and converges to Ω.

Theorem 5.13. Whenever $\omega_n > \Omega_i$, then
$$\Omega_i < \omega_n < \Omega \leq \Omega_i + 2^{-i}.$$

Given any i, if we know the first i bits of Ω, we can decide the halting of any program with length $\leq i$.

Proof. The first sentence is an immediate consequence of the inequality

$$2^{-i} \geq \sum_{j=i+1}^{\infty} b_j 2^{-j}.$$

To prove the second sentence, assume that we know B_i. Hence, we are able to compute Ω_i. We now compute the numbers ω_j until we have found an n such that $\omega_n > \Omega_i$. By the properties of ω_j, this is always possible.

Let u_1 be a word over V of length $i_1 \leq i$. We claim that $U(u_1, \lambda)$ is defined iff u_1 is one of the words $g(1), \ldots, g(n)$. The "if"-part of the claim is clear. To prove the "only if"-part, we assume the contrary: $u_1 = g(m)$, where $m > n$. We obtain a contradiction by the following chain of inequalities:

$$\Omega > \omega_m \geq \omega_n + 2^{-i_1} \geq \omega_n + 2^{-i} > \Omega_i + 2^{-i} \geq \Omega.$$

□

Theorem 5.13 justifies the term "magic bits". The knowledge of a sufficiently long prefix B_i enables us to solve any halting problem, and i can be computed from the halting problem. The same applies to Post Correspondence Problems as well, because we can design programs to search for a solution of a given PCP. Similarly, we can design programs to look for counterexamples of famous conjectures in classical mathematics, such as Fermat's last problem or Riemann's hypothesis. We have already indicated at the beginning of Chapter 5 how the knowledge of a sufficiently long sequence B_i of magic bits opens even more general vistas. It enables us to decide whether a well-formed formula is, according to a formal theory, a theorem, a nontheorem or independent.

How many magic bits are actually needed depends of course on the formal theory, but also on the programming of the universal computer U. Certainly our programming of U in the proof of Theorem 5.2 was not very economical in this respect. We started the programs with $0^i 1$, where i is the index of an individual computer C_i. A much more compact programming for U results if we take the programs of the individual computers C_i as such and use the technique of Example 5.4 to make words self-delimiting. Then for any conceivably interesting formal theories the knowledge of 10 000 magic bits, that is B_i with $i = 10\,000$, is more than enough.

Thus, if an oracle tells us 10 000 magic bits, we are wise enough to settle all halting problems and PCPs that are of reasonable size, as well as famous conjectures in classical mathematics. Although we are wise in this sense, we still face the task of enormous computational complexity when we start calculating the numbers ω_n.

We will now prove that Ω is truly random.

Theorem 5.14. *The sequence B is random.*

Proof. We apply the notation from the proof of Theorem 5.13. We showed that whenever $U(u_1, \lambda)$ is defined and $|u_1| \leq i$, then u_1 is one of the words $g(1), \ldots, g(n)$.

This leads to the equation

(∗) $\qquad \{U(g(j),\lambda)|1\leq j\leq n \text{ and } |g(j)|\leq i\} = \{w|H(w)\leq i\}$.

Indeed, every word belonging to the left side belongs to the right side and, conversely, if $H(w) \leq i$ then w has a program of length at most i.

Consider now the partial recursive function f from V^* into V^*, defined as follows. Given $x = x_1 \ldots x_t$, $x_j \in V$, find the smallest m, if any, such that

$$\omega_m > \sum_{j=1}^{t} x_j 2^{-j}.$$

If such an m is found, $f(x)$ is the first word (in lexicographical order) not belonging to the set

$$\{g(j)|1 \leq j \leq m\}.$$

Let C be the computer defined by

$$C(x,\lambda) = f(U(x,\lambda)).$$

We now consider an arbitrary prefix B_i and deduce

$$H(f(B_i)) \leq H_C(f(B_i)) + \text{sim}(C)$$

$$= \min\{|u| \mid C(u,\lambda) = f(B_i)\} + \text{sim}(C)$$

$$= \min\{|u| \mid f(U(u,\lambda)) = f(B_i)\} + \text{sim}(C)$$

$$\leq \min\{|u| \mid U(u,\lambda) = B_i\} + \text{sim}(C)$$

$$= H(B_i) + \text{sim}(C).$$

By (∗), $H(f(B_i)) > i$. Hence,

$$i - \text{sim}(C) < H(B_i)$$

for all i, showing that B is random.

□

It is clear by the discussion following Theorem 5.13 that an upper bound n is inherent in every formal theory, such that no prefix B_i with $i > n$ can be produced according to the theory. Theorem 5.14 shows that such an upper bound also concerns the total number of bits of Ω that can be produced according to the theory. Essentially, if some theory could produce infinitely many magic bits, even if they were not consecutive bits in B, the contribution of the theory would exceed any constant for sufficiently long prefixes B_i. Intuitively, a gambler could use the theory for infinite gain.

We are now ready to take the final step. A really dramatic implication of the properties of Ω is that we can exhibit a particular (exponential) Diophantine equation

(*) $$P(i, x_1, \ldots, x_m) = Q(i, x_1, \ldots, x_m)$$

such that, for each i, (*) has infinitely many solutions in $x_1, \ldots, x_m$ iff b_i, the ith bit in Ω, equals 1. As in the definition of an exponential Diophantine relation, P and Q are functions built up from $i, x_1, \ldots, x_m$ and nonnegative integer constants by the operations of addition, multiplication and exponentiation. Denote by $(*)_i$, $i = 1, 2, \ldots$, the equation obtained from (*) by fixing the parameter i. Does $(*)_i$ have infinitely many solutions? By the properties of Ω deduced above, any formal theory can answer this question for finitely many values of i only. No matter how many additional answers we obtain, for instance by experimental methods or just flipping a coin, this won't help us in any way as regards the remaining infinitely many values of i. As regards these values, mathematical reasoning is helpless, and a mathematician is no better off than a gambler flipping a coin. This holds in spite of the fact that we are dealing with basic arithmetic.

In fact, we have already developed all the technical apparatus needed to establish the above claim concerning (*) and the magic bits. Recall the definition

$$\omega_n = \sum_{j=1}^{n} 2^{-|g(j)|},$$

where g constitutes an enumeration of programs for U that halt with the empty word. We fix an i and consider the ith bit $b(i, n)$ in ω_n for increasing values of n. At first $b(i, n)$ may fluctuate irregularly between 0 and 1 but it will stabilize from a certain point onwards. The reason for this is that because ω_n tends to Ω, $b(i, n)$ has to become b_i.

The binary relation $b(i, x_1) = 1$ is recursively enumerable — in fact, it is recursive. (We have changed the notation from n to x_1 to conform with (*).) As shown in Section 3.1, there are P and Q such that (*) has a solution in $x_2, \ldots, x_m$ iff $b(i, x_1) = 1$. Moreover, this ED representation is *singlefold*: for each i and x_1, there is at most one $(m-1)$-tuple $(x_2, \ldots, x_m)$ satisfying (*). Consequently, for each i, $b(i, x_1) = 1$ holds for infinitely many values of x_1 iff (*) holds for infinitely many m-tuples $(x_1, \ldots, x_m)$. Because $b_i = 1$ iff $b(i, x_1) = 1$ for infinitely many values of x_1, we have established the desired result.

Theorem 5.15. *The ith magic bit b_i equals 1 iff (*) has infinitely many solutions.*

Chaitin [Cha 2] has constructed (*) explicitly. His equation is really huge: some 17 000 variables and 900 000 characters.

It is essential that we ask whether or not (*) has *infinitely many* solutions for a given i. It is not sufficient to consider the *existence* of solutions because the set of (the indices of) solvable Diophantine equations is recursively enumerable and, hence, cannot lead to randomness.

Singlefoldedness could be replaced by a weaker property: for each i and x_1, $(*)$ holds for only finitely many $(m-1)$-tuples $x_2, \ldots, x_m$. It is not known whether the theory developed in Section 3.2 holds for this weaker notion of singlefoldedness and, thus, it is also not known whether P and Q in $(*)$ can be assumed to be polynomials.

Epilogue

Tarzan. I would like to speak about some of my views and observations, since I have now read through all five chapters. Please tell me if I say something directly wrong.

When you speak about problems with many faces, you merely show how an instance of one "face" can effectively be transformed into an instance of another "face". It is like transforming one Turing machine into another. The transformation is between imaginary machines because, in the case of undecidability, no machines can exist.

A discrete parameter, an integer, usually estimates borderlines only in a very rough way. If you take steps of a thousand kilometres to approach a borderline in nature, your step is not likely to come close to it. To approach the borderline between decidable and undecidable, you want to make the scope of instances of your decidable problem larger and larger. You should then use different parameters. When changing one parameter to another, you may lose on some front but gain on another. Comparative equivalence problems between language families discussed in Section 4.2 offer good examples.

Zero-knowledge proofs are different from the rest of the topics. Here you really talk about *efficiency* rather than *effectiveness*. You do not gain any unwanted knowledge during some specified time which can be your lifetime. Given an unlimited amount of time, you are able to unlock all boxes.

Bolgani. This applies to cryptography in general.

Tarzan. The direct undecidability arguments, that is, the ones not based on reductions, are diagonal in flavour. Self-referential. We feed a Turing machine its own description and see what happens. What can be said about self-reference in formal systems in general?

Kurt. We have seen that in a formal system we can construct statements about the formal system, of which some can be proved and some cannot, according to what they say about the system. We shall compare this fact with the famous

Epimenides paradox ("Der Lügner"). Suppose that on 4 May 1934, A makes the single statement, "Every statement which A makes on 4 May 1934 is false." This statement clearly cannot be true. Also it cannot be false, since the only way for it to be false is for A to have made a true statement in the time specified and in that time he made only the single statement.

The solution suggested by Whitehead and Russell, that a proposition cannot say something about itself, is too drastic. We saw that we can construct propositions which make statements about themselves and, in fact, these are arithmetic propositions which involve only recursively defined functions, and therefore are undoubtedly meaningful statements. It is even possible, for any metamathematical property f which can be expressed in the system, to construct a proposition which says of itself that it has this property. For suppose that $F(z_n)$ means that n is the number of a formula that has the property f. Then, if $F(S(w,w))$ has the number p, $F(S(z_p, z_p))$ says that it has the property f itself. This construction can only be carried out if the property f can be expressed in the system, and the solution of the Epimenides paradox lies in the fact that the latter is not possible for every metamathematical property. For, consider the above statement made by A. A must specify a language B and say that every statement that he made in the given time was a false statement in B. But "false statement in B" cannot be expressed in B, and so his statement was in some other language, and the paradox disappears.

The paradox can be considered as a proof that "false statement in B" cannot be expressed in B.

Tarzan. Such an analysis would be impossible without considering formal systems, without the axiomatic method.

David. The axiomatic method is in fact an unavoidable tool for all exact research, no matter what the research area is. It is suitable for the human mind, logically well founded and fruitful. It gives to the research full freedom of movement. To proceed axiomatically means to think with full awareness. Before the axiomatic method one believed in certain things as in dogmas. This naïve belief is removed by the axiomatic method; still the earlier advantages remain. The principles of argument in mathematics become fully clear. However, we can never be sure *a priori* about the consistency of our axioms.

Tarzan. Still, a lot of things escape any given axiom system. Let us go to matters discussed in Chapter 5. It is difficult for me to visualize that we just flip a coin to decide whether an equation has a finite or an infinite number of solutions. Both outcomes are OK and in accordance with any theory we might have had before. And the additional knowledge gained by deciding about this particular equation does not help us much: infinitely many equations remain to be handled similarly.

Bolgani. You might have difficulties only because you forget that everything can be encoded as solutions of equations. This was done in Chapter 3. We might equally well ask whether a given Turing machine defines a finite or an infinite language. It might be easier for you to visualize that no formal system can answer this question in regard to all Turing machines. There always will be machines for which the decision has to be made by coin flipping. After all, a formal system itself is nothing but a Turing machine.

Tarzan. I can follow all that. Still, we have a specific equation. Even if it is long, it can be written down explicitly, and it has been written down explicitly. The question whether it has a finite or an infinite number of solutions is clear enough. In my world, call it Platonic if you like, this question has a definite answer. It is only in my ignorance that I don't know the answer. In my world the law of the excluded middle holds in generality, at least in arithmetical matters like this. I like the idea that we always discover new things from this world, never being able to exhaust it.

Bolgani. Think of the infinite sequence of equations about which we have to make the decision: finitely or infinitely many solutions, 0 or 1. So we obtain an infinite sequence of 0s and 1s for each completed set of decisions. Each such sequence corresponds to a (hopefully) consistent theory. There are many sequences, even if we take into account that some of them do not correspond to an axiomatizable theory (this happens if the sequence is not recursive). Which sequence is the one of your Platonic world? Is your world the same as mine? You also admitted that you don't know your world so well. Finding the answers is discovery, not creation. Objects of set theory should be discovered. You can also encode things in different ways, symbolize, as Post says. Here the physical, experimental aspect is important: which objects correspond to the universe we live in? Which of the Platonic worlds is realized, is a question of physics.

Emil. The creativeness of human mathematics has as its counterpart inescapable limitations thereof — witness the absolutely unsolvable problems. The phrase "absolutely unsolvable" is due to Church, who thus described the problem in answer to my query as to whether the unsolvability of his elementary number theory problem was relative to a given logic. A fundamental problem is the existence of absolutely undecidable propositions, that is, propositions which in some *a priori* fashion can be said to have a determined truth-value, and yet cannot be proved or disproved by any valid logic. I cannot overemphasize the fundamental importance to mathematics of the existence of absolutely unsolvable problems. True, with a specific criterion of solvability under consideration, the unsolvability becomes merely unsolvability by a given set of instruments (as in the case of the famous problems of antiquity). The fundamental new thing is that the given set of instruments is the only humanly possible set.

Mathematical thinking is, and must be, essentially creative. It is to my continuing amazement that ten years after Gödel's remarkable achievement views on the nature of mathematics were thereby affected only to the point of seeing the need for many formal systems, instead of a universal one. Rather, has it seemed to us to be inevitable that these developments will result in a reversal of the entire axiomatic trend of the late nineteenth and early twentieth centuries, with a return to meaning and truth. Postulational thinking will then remain but one phase of mathematical thinking.

Tarzan. The rest is silence. I have to go.

Bibliography

[Bör] E. Börger, *Computability, Complexity, Logic*. North-Holland Publishing Company, Amsterdam (1989).

[Cal] C. Calude, *Information and Randomness, an Algorithmic Perspective*. In preparation.

[Cha 1] G. Chaitin, A theory of program size formally identical to information theory. *Journal of the Association for Computing Machinery* 22 (1975) 329–340.

[Cha 2] G. Chaitin, *Algorithmic Information Theory*. Cambridge University Press, Cambridge (1987).

[Chu] A. Church, An unsolvable problem of elementary number theory. *The American Journal of Mathematics* 58 (1936) 345–363.

[Dav] M. Davis (ed.), *The Undecidable*. Raven Press, Hewlett, N.Y. (1965).

[Göd 1] K. Gödel, Über formal unentscheidbare Sätze der Principia Mathematica und verwandter Systeme I. *Monatshefte für Mathematik und Physik* 38 (1931) 173–198.

[Göd 2] K. Gödel, *Collected Works*, vol. 1. (S. Feferman, J. Dawson, S. Kleene, G. Moore, R. Solovay, and J. van Heijenvort, eds.), Oxford University Press, Oxford and New York (1986).

[Hil] *Hilbertiana. Fünf Aufsätze von David Hilbert*. Wissenschaftliche Buchgesellschaft, Darmstadt (1964).

[Ho Ul] J. Hopcroft and J. Ullman, *Introduction to Automata Theory, Languages and Computation*. Addison-Wesley, Reading, Mass. (1979).

[Jo Ma] J. Jones and Ju. Matijasevitch, Register machine proof of the theorem on exponential diophantine representation of enumerable sets. *Journal of Symbolic Logic* 49 (1984) 818–829.

[Ku Sa] W. Kuich and A. Salomaa, *Semirings, Automata, Languages*. Springer-Verlag, Berlin, Heidelberg, New York (1986).

[Li Vi] M. Li and P. Vitanyi, Kolmogorov complexity and its applications. In: J. van Leeuwen (ed.) *Handbook of Theoretical Computer Science*, Vol. A (1990) 187–254.

[Min] M. Minsky, *Computation: Finite and Infinite Machines*. Prentice-Hall, Englewood Cliffs, N.J. (1967).

[Post 1] E. Post, Finite combinatory processes – Formulation 1. *Journal of Symbolic Logic* 1 (1936) 103–105.

[Post 2] E. Post, A variant of a recursively unsolvable problem. *Bulletin of the American Mathematical Society* 52 (1946).

[Post 3] E. Post, Formal reductions of the general combinatorial decision problem. *American Journal of Mathematics* 65 (1943) 197–215.

[Pr Li] P. Prusinkiewicz and A. Lindenmayer, *The Algorithmic Beauty of Plants*. Springer-Verlag, Berlin, Heidelberg, New York (1990).

[Rog] H. Rogers, *Theory of Recursive Functions and Effective Computability*. McGraw-Hill, New York (1967).

[Ro Sa] G. Rozenberg and A. Salomaa, *The Mathematical Theory of L Systems*. Academic Press, New York (1980).

[Sal 1] A. Salomaa, *Formal Languages*. Academic Press, New York (1973).

[Sal 2] A. Salomaa, *Computation and Automata*. Cambridge University Press, Cambridge (1985).

[Sa So] A. Salomaa and M. Soittola, *Automata-Theoretic Aspects of Formal Power Series*. Springer-Verlag, Berlin, Heidelberg, New York (1978).

[Thue] A. Thue, Über unendliche Zeichenreihen. *Skrifter utgit av Videnskapsselskapet i Kristiania* I (1906) 1–22.

[Tur] A. Turing, On computable numbers, with an application to the Entscheidungsproblem. *Proceedings of the London Mathematical Society* (2), 42 (1937) 230–265.

Index

acceptable, 20
algorithm, 1
algorithmic entropy, 169
algorithmic information, 174
alphabet, 25

blank, 4
busy beaver, 7

canonical program, 168
Cantor's diagonal argument, xiii
catenation, 25
catenation closure, 26
Chaitin complexity, 175
Champernowne's number, 163
characteristic function, 3
characteristic series, 119
Chomsky hierarchy, 34
Church's Thesis, 9
circular variant, 35
complete, 42
compressibility, 162, 169
computable, 1
computation, 6
computational complexity, 22
computer, 165
conditional complexity, 169
consistent, 42
context-free, 34
context-sensitive, 34

creative, 40, 42

Davis–Putnam–Robinson Theorem, 93
decidable, 1, 6
decidable problem, 2
decision problem, xiv
definable, 20
degree, 109
degree of ambiguity, 80
degree of randomness, 183
degrees of undecidability, 39
deletion language, 151
derivation, 29
descriptional complexity, 175
deterministic, 5
dilemma of diagonalization, 13
dimension, 109
Diophantine, 92
discrete logarithm, 155
DOL system, 36, 140
dominance relation, 96
DTOL function, 124
dyadic, 3

ED, 93
effective procedure, 1
Ehrenfeucht's Conjecture, 72
empty word, 25
equality set, 65
equivalent, 29

exponential Diophantine, 92

Fibonacci sequence, 120
final state, 5
finite automaton, 34
formalism, xiii
formal language, 26
formal power series, 118
frequency indicator, 164

Generalized Post Correspondence Problem, 61
Gödel–Rosser Incompleteness Theorem, 44
Gödel number, 10, 42
Gödel sentence, 44
Gödel's Second Incompleteness Theorem, 45
grammar, 29

Hadamard product, 118
halting probability, 181
halting problem, 11
Hilbert's Tenth Problem, 91

incompleteness, xiv
index, 10, 42
inductive inference, 163
initial state, 5
instantaneous description, 6
instruction, 5
intractable, 23
Invariance Theorem, 169

Kolmogorov complexity, 176
Kraft inequality, 164

L system, 36
letter, 25
Lindenmayer system, 36
linear language, 35
literal, 157
logicism, xiii
Lyndon Theorem, 25

magic bits, 184
Matijasevitch Theorem, 110
matrix grammar, 137

membership problem, 148
mirror image, 35
Modified Post Correspondence Problem, 62
morphism, 26

noncomputable, 4
nondeterministic Turing machine, 23
nonerasing, 26
noninteractive, 160
nonterminal, 29
NP, 24
NP-complete, 24
NP-hard, 24

0L systems, 36
one-way function, 23

P, 23
pairing function, 11
Parikh equivalent, 35
Parikh vector, 35
partial function, 1
partial recursive, 6
pattern language, 149
Peano arithmetic, 42
Pell-equation, 112
periodic, 59
Petri net, 139
polynomially bounded, 23
Post canonical system, 36
Post Correspondence Problem, 50, 51
Post normal system, 37
prefix-free, 164
primitive, 50
production, 28
productive, 40, 42
program tape, 166
properly thin, 145
pushdown automaton, 34
pushdown stack, 34

reachability problem, 135
recursion theorem, 37
recursive, 31
recursively enumerable, 14, 31, 42
recursively inseparable, 42

Index

recursiveness, 4
recursive, 6
register machine, 17, 98
regular, 34
regular expression, 34
rewriting rule, 28
rewriting system, 25
Rice's Theorem, 39

satisfiability problem, 24, 156
self-applicability problem, 13
self-delimiting, 177
semilinear, 36
semi-Thue problem, 32, 146
semi-Thue system, 28
sequence equivalence, 140
Sieve of Eratosthenes, 2
simulation constant, 167
singlefold, 92
slender, 81
solvable, 4
space complexity, 24
state, 4
subword, 25
support, 119

tape, 4
terminal, 29
test set, 70
thin, 81
Thue problem, 32, 146
Thue system, 29
time complexity, 22
TOL system, 36
total function, 1
total recursive, 6
Turing machine, 4, 6, 30

unambiguous, 35
unary notation, 5
undecidable, 4
universal computer, 168
universal polynomial, 110
universal Turing machine, 14
UOL system, 134

USF language, 133

vector addition system, 135

ω-consistent, 45

well-formed formula, 41
word, 25
word equation, 72
word problem, 32, 147
work tape, 166

yield relation, 28

Z-rational, 119
zero-knowledge, 153